Biostimulants in Plant Science

Edited by Seyed Mahyar Mirmajlessi and Ramalingam Radhakrishnan

Published in London, United Kingdom

IntechOpen

Supporting open minds since 2005

Biostimulants in Plant Science
http://dx.doi.org/10.5772/intechopen.82916
Edited by Seyed Mahyar Mirmajlessi and Ramalingam Radhakrishnan

Contributors
Akinlolu Olalekan Akanmu, Barachel Odaro-Junior Umukoro, Kanayo Stephen Chuckwuka, Micheal Dare Asemoloye, Adegboyega Christopher Odebode, José Alfonso Domínguez-Núñez, Ada S. Albanesi, Pradeep Kumar Singh, Meenakshi Singh, Manjari Mishra, Devendra Kumar Srivastava, Ana Carolina Feitosa De Vasconcelos, Lúcia Helena Garófalo Chaves, Sovan Debnath, Deepa Rawat, Aritra Kumar Mukherjee, Samrat Adhikary, Ritesh Kundu, Yuvaraj Muthuraman, Murugaragavan Ramasamy, Elavarasi Prabakaran, Shishir Raut, Saurabh Kumar Singh

First published in London, United Kingdom, 2020 by IntechOpen
IntechOpen is the global imprint of INTECHOPEN LIMITED, registered in England and Wales, registration number: 11086078, 7th floor, 10 Lower Thames Street, London, EC3R 6AF, United Kingdom
Printed in Croatia

British Library Cataloguing-in-Publication Data
A catalogue record for this book is available from the British Library

Additional hard and PDF copies can be obtained from orders@intechopen.com

Biostimulants in Plant Science
Edited by Seyed Mahyar Mirmajlessi and Ramalingam Radhakrishnan
p. cm.
Print ISBN 978-1-83880-161-8
Online ISBN 978-1-83880-162-5
eBook (PDF) ISBN 978-1-83880-413-8

For EU product safety concerns:
IN TECH d.o.o., Prolaz Marije Krucifikse Kozulić 3, 51000 Rijeka, Croatia, info@intechopen.com or visit our website at intechopen.com.

Meet the editors

Dr. Seyed Mahyar Mirmajlessi is a highly experienced plant pathologist with particular expertise related to disease management, plant-pathogen interactions, and biological control. He received his MSc degree in plant pathology with a specialization in genetic diversity of plant-pathogenic fungi. He earned his Ph.D. degree in molecular plant pathology from the Estonian University of Life Sciences. His career continued as a postdoctoral researcher in plant protection at Ghent University, Belgium. Dr. Mirmajlessi has acted as reviewer and editorial board member for a number of scientific journals published by Elsevier, Springer, APS, and ACS. He has also published several research articles in various international peer-reviewed journals. Currently, he is working as senior research associate in the Department of Plants and Crops, Faculty of Bioscience Engineering, Ghent University, Belgium.

Dr. Ramalingam Radhakrishnan was born in India. He has received several research awards and fellowships during his Doctor of Philosophy studies and he has made a significant contribution in 'Application of magnetic field on improvement of crop plants'. His research was honored by the Chinese Academy of Science providing financial support to present his findings in an international conference held in China. Professionally, he has been employed as a post-doctoral researcher, research professor, and assistant professor in India and South Korean universities and research institutes. His major research finding is the utilization of microbes or elicitors to improve the crop under environmental stress conditions and biological weed control. He has published several research and review papers as main author in reputed journals, books, and conferences.

Contents

Preface

Over the past few years, natural-based crop production substances, 'plant biostimulants', have been considered as environmentally friendly alternatives to agrichemicals. A plant biostimulant describes any microorganism and/or substance applied to seeds, plants, or soil microbiota to promote plant physiological pathways such as mineral nutrient uptake, crop quality, growth regulation, and tolerance to unfavorable environmental conditions. Nowadays, a large number of biostimulants are found as a complex chemical mixture originated from biological processes with plant-nutritive elements as bioinoculants enhancing nutrient availability to plants. In this sense, biostimulants may comprise fungal or bacterial inoculants, humic acids, amino acids, fulvic acids, seaweed extracts, etc. Biostimulants have biopesticide and biostimulant utilities affecting their regulatory classification.

Elucidations on direct or microbially mediated functions of biostimulants are presented in this book to illustrate fundamental principles, modes of action, and recent applications underlying this technology. The papers selected for this book comprise a cross-section of topics that reflect an overview on concepts of plant biostimulants from different points of view in order to describe effective strategies for using these substances and/or beneficial microorganisms within sustainable agroecosystems. This book, "Biostimulants in Plant Science", encompasses two main sections covering nine reviews focused on 'Elucidation of biostimulant functions on promoting plant growth', and 'The role of bacterial and fungal communities on enhancing nutrient bioavailability'. I hope that these chapters adequately reflect the objectives of this compilation.

Elucidation of Biostimulant Functions on Promoting Plant Growth

The first chapter, "Biostimulants and Their Role in Improving Plant Growth under Abiotic Stresses", discusses the use of biostimulants in plant growth according to the raw material used in their compositions as well as their effects on plants subjected to abiotic stresses.

The second chapter, "Application of Bacteria as a Prominent Source of Biofertilizers", provides an overview of different bacterial biofertilizers and its associations with plants and nutrients transformations in soil. This chapter adopts a rational approach to use for the management of microbial fertilizers in sustainable agriculture and it has vast potential for the future.

The third chapter, "Applications and Constraints of Plant Beneficial Microorganisms in Agriculture", introduces biofertilizers as highly potent alternatives to inorganic fertilizers and also as an economically attractive route for augmenting nutrient supply. Beneficial microorganisms have the potential ability to fix atmospheric nitrogen, solubilize and mobilize plant nutrients from the insoluble form through a microbiological process.

The fourth chapter, "Biochar: A Vital Source for Sustainable Agriculture", reviews the contributions of biochar technology to environmental sustainability and food security. This strategy addresses the declining food security issues, depleting soil, and plant health challenges. Biochar enhances biological nitrogen fixation and productivity by facilitating changes in plants' physical conditions, rapid germination, and growth. It also enhances nutrient uptake, water holding capacity, and microbial activity or acts against biotic and abiotic stresses.

The fifth chapter, "Role of Soil Microbes on Crop Yield against Edaphic Factors of Soil", focuses on losing soil productivity where continuous usage of inorganic fertilizers coupled with depletion of organic matter results in deterioration of soil structure and its productivity. It also leads to a reduced input/output ratio unless soils are replenished with organic matter through green manure, compost, or microbial activity. Thus, the microbes can be utilized to overcome the harmful effect of chemical degradation of soil and waterlogging, which improves soil fertility.

The Role of Bacterial and Fungal Communities on Enhancing Nutrient Bioavailability

The sixth chapter, "Role of Fungi in Agriculture", firstly gives a brief introduction on fungal filaments that enhance the nutrient availability by solubilizing insoluble nutrients like phosphorus, and increase the nutrient mobility due to faster intracellular nutrient mobility. Then, the arbuscular mycorrhizal fungi (AMF) are introduced for how to protect plants by up-regulating the activity of antioxidant enzymes and osmolytes, and by regulating the synthesis of phytohormones, which might possibly interconnect the various tolerance mechanisms for cumulative stress response.

The seventh chapter, "Arbuscular Mycorrhiza-Associated Rhizobacteria and Biocontrol of Soilborne Phytopathogens", provides a general picture on understanding the mechanisms involved in arbuscular mycorrhizal fungi (AMF)/rhizobacteria interactions including the mechanisms of AMF-mediated biocontrol; interactions between AMF associated bacteria (AMB) and extraradical mycelium network of AMF; AM associated bacteria and biocontrol activities; and the unfavorable zone to pathogen development as the mycorrhizosphere.

The eighth chapter, "Ectomycorrhizal Fungi as Biofertilizers in Forestry", describes the value of ECM fungi from a global framework, not only to increase the production of edible fruit bodies but also for the regular practices of reforestation and restoration of ecosystems with implicit applications in biofertilization, bioremediation, and control of soil pathogens. Moreover, ecological functions, the direct implications of the ECM fungi as biofertilizers in forest management are briefly discussed: reforestation, plantation management, and ecosystem restoration.

The ninth chapter, "Microbes for Iron Chlorosis Remediation in Peach", addresses the current trend of detection methods and control measures of iron chlorosis in peaches, and gives attention to bioremediation techniques for the correction of lime-induced iron chlorosis. Traditional soil and foliar application methods including ferrous sulphate, Fe-EDTA, Fe-EDDHA chelates, etc. cannot be considered as reliable corrective measures of chlorosis. Besides, the importance of microbe-mediated correction strategies in iron fixation in calcareous soil and iron uptake by plants is discussed.

IntechOpen has taken a commendable step to publish a series of valuable books in the context of plant sciences. So, it is with great pleasure that this book has attracted attention from researchers who were selected based on their previous contributions in scientific journals. I sincerely hope that the materials of this book will help a wide range of readers to update their insights on the role of biostimulants in plant science, and particularly in sustainable agriculture. Therefore, it can be useful especially when you are beginning your career or are just researching this topic. Last but not least, I would like to thank IntechOpen for inviting me to be the book editor. Special thanks also go to Ms. Rebekah Pribetic, Author Service Manager, for her help and cooperation during the whole editing process.

Dr. Ir. Seyed Mahyar Mirmajlessi
Department of Plants and Crops,
Faculty of Bioscience Engineering,
Ghent University (UGent),
Belgium

Ramalingam Radhakrishnan
Jamal Mohamed College (affiliated to Bharathidasan University),
India

Section 1

Elucidation of Biostimulant Functions on Promoting Plant Growth

Chapter 1

Biostimulants and Their Role in Improving Plant Growth under Abiotic Stresses

Ana Carolina Feitosa de Vasconcelos and Lúcia Helena Garófalo Chaves

Abstract

Biostimulants are products that reduce the need for fertilizers and increase plant growth, resistance to water and abiotic stresses. In small concentrations, these substances are efficient, favoring the good performance of the plant's vital processes, and allowing high yields and good quality products. In addition, biostimulants applied to plants enhance nutrition efficiency, abiotic stress tolerance and/or plant quality traits, regardless of its nutrient contents. Several researches have been developed in order to evaluate the biostimulants in improving plant development subjected to stresses, saline environment, and development of seedlings, among others. Furthermore, various raw materials have been used in biostimulant compositions, such as humic acids, hormones, algae extracts, and plant growth-promoting bacteria. In this sense, this chapter aims to approach the use of biostimulants in plant growth according to the raw material used in their compositions as well as their effects on plants subjected to abiotic stresses.

Keywords: drought, salinity, temperature, humic substances, seaweed extracts, hormones, arbuscular mycorrhizal fungi, plant growth-promoting rhizobacteria

1. Introduction

Biostimulants are natural or synthetic substances that can be applied to seeds, plants, and soil. These substances cause changes in vital and structural processes in order to influence plant growth through improved tolerance to abiotic stresses and increase seed and/or grain yield and quality. In addition, biostimulants reduce the need for fertilizers [1].

Many definitions of biostimulants have been reported [2]. According to [3], biostimulants could be classified depending on the mode of action and the origin of the active ingredient; while Ref. [4] proposed biostimulants should be classified based on their action in the plants or, on the physiological plant responses rather than on their composition. In addition Ref. [1] has emphasized the importance of the final impact on plant productivity which suggests that any definition of biostimulants should focus on the agricultural functions of biostimulants, either on the nature of their constituents or on their modes of actions.

Thus Ref. [2] proposed the following definition of a biostimulant as a formulated product of biological origin that improves plant productivity because of the novel or emergent properties of the complex of constituents; and not as a sole consequences of the presence of known essential plant nutrients, plant growth regulators, or plant protective compounds. This definition is important as it emphasizes the principle that biological function can be positively modulated through the application of molecules, or mixtures of molecules, for which an explicit mode of action has not been defined.

In small concentrations, these substances are efficient, enhancing nutrition efficiency, abiotic stress tolerance, and/or crop quality traits, regardless of its nutrients content. These substances when applied exogenously have similar actions to the groups of known plant hormones, whose main ones are auxins, gibberellins, and cytokinins [5].

Abiotic stress is a problem of concern for the growth and productivity of plants in modern times. Abiotic stresses, such as drought, salinity, and extreme temperatures, are responsible for huge crop losses globally [6]. In order to prevent these losses, biostimulants are increasingly being integrated into production systems with the goal of modifying physiological processes in plants to optimize productivity [2].

In general, biostimulants are produced as a junction of natural or synthetic substances composed of hormones or precursors of plant hormones. When applied correctly in the crops, it acts directly on the physiological processes providing potential benefits for growth, development, and/or responses to water stress, saline, and toxic elements, such as toxic aluminum [7, 8].

These products, which differ from traditional nitrogen, phosphorus, and potassium fertilizers, may contain in their formula a variety of organic compounds, such as humic acids, seaweed extracts, vitamins, amino acids, ascorbic acid, and other chemicals, which may vary according to its manufacturer [5].

Biostimulants offer a potentially novel approach for the regulation and/or modification of physiological processes in plants to stimulate growth, to mitigate stress-induced limitations, and to increase yield. The effects of biostimulants are still not clear. They can act on plant productivity as a direct response of plants or soils to the biostimulant application or an indirect response of the biostimulant on the soil and plant microbiome with subsequent effects on plant productivity [2].

Several researches have been developed in order to evaluate the use of biostimulants in improving plant growth subjected to abiotic stresses. Furthermore, various raw materials have been used in biostimulant compositions, such as humic acids, hormones, algae extracts, and plant growth-promoting bacteria [7].

In this sense, this chapter aims to approach the use of biostimulants in crops under abiotic stresses and their effects on plant growth.

2. Biostimulants and abiotic stresses in plants

Abiotic stress is defined as environmental conditions that reduce growth and yield below optimum levels [9]. Abiotic stress such as cold, drought, and salt largely influences plant development and crop productivity. Abiotic stress has been becoming a major threat to food security due to the constant changes in climate and deterioration of the environment caused by human activity. To cope with abiotic stress, plants can initiate a number of molecular, cellular, and physiological changes to respond and adapt to such stresses [10].

Abiotic stresses may be prevented by optimizing plant growth conditions and through the provision of water and nutrients and plant growth regulators (PGRs—auxins, cytokinins, gibberellins, strigolactones, and brassinosteroids). In addition to these traditional approaches, biostimulants have been highlighted as a promoter of optimizing productivity by modifying physiological processes in plants. Biostimulants offer a potentially novel approach for the regulation and/or modification of physiological processes in plants to stimulate growth, to mitigate stress-induced limitations, and to increase yield [2].

The plant hormone auxin is the key regulator of many aspects of plant growth and development, including cell division and stretching, differentiation, tropisms, apical dominance, senescence, abscission, and flowering. The cytokinins are mainly responsible for cell division, besides affecting many other processes, such as vascular development, apical dominance, and nutrient mobilization, especially when interacting with auxins [11].

Gibberellic acid has a marked effect on the seed germination process, activating hydrolytic enzymes, such as α-amylase and protease, which actively act in the unfolding of the reserve substances, facilitating the mobilization of the endosperm. In addition, they promote the breakdown of dormancy, stem elongation and growth, cell division, and, consequently, leaf expansion [12].

According to Ref. [13], the biostimulant is composed of cytokinin, indolebutyric acid, and gibberellic acid, applied in seed, increased the seedling emergence percentage of *Gossypium hirsutum* L., as well as leaf area, height, and growth of seedlings. The algal extract applied via leaf yielded higher seed yield of *Glycine max* (L.) Merr [14].

An increase in the quantity and quality of *Allium cepa* L. bulbs with foliar application of putrescine and amino acid glutamine was observed [15]. L-glutamic acid is an important amino acid that acts as a central molecule in the metabolism of higher plants [16], being the precursor of the synthesis of chlorophyll in leaves [5], and the carbon regulatory function and nitrogen metabolism [17]. Glutamate is also a precursor of arginine and ornithine, which in turn act on the synthesis of polyamines, which can act on plants, minimizing stress conditions [18, 19]. In addition to these amino acids, others are important in cell metabolism with the expressive diversity of biological functions.

The application of extracts from algae or other plants have beneficial effects on growth and stress adaptation. Algal extracts, protein hydrolysates, humic and fulvic acids, and other compounded mixtures have properties beyond basic nutrition, often enhancing growth and stress tolerance. Although most plant biostimulants are added to the rhizosphere to facilitate uptake of nutrients, many of these also have protective effects against environmental stress such as water deficit, soil salinization, and exposure to sub-optimal growth temperatures [20].

2.1 Biostimulants and water stress in plants

Drought is one of the most important and prevalent stress factors for plants in many parts of the world, especially in arid and semiarid areas. Drought stress is a multidimensional stress and generally leads to changes in the physiological, morphological, ecological, biochemical, and molecular traits of plants. In addition, it can negatively affect the quantity and quality of plant growth and yield. Plants respond to water deficit depending on the length and severity of the water deficiency as well as the plant species, age, and developmental stage [21].

Biostimulants when applied to seeds or early plant development stimulate root production and growth [22], especially in soils with low fertility and low water availability, acting on the accelerated recovery of the seedlings in unfavorable conditions, such as water deficit. These products, especially the organic ones, reduce the need of fertilizers to the plants, and increase their productivity and resistance to water and climatic stress, since they act as a hormonal and nutritional increment [23].

Consequently, a series of biostimulants were developed and marketed mainly in the agricultural sector. For example, biostimulants marketed under the trade names Generate, Crop Set, Fulcrum, and Redicrop 2000 worked positively in both the root system and leaf spray in three tree species (*Quercus rubra, Betula pendula*, and *Fagus sylvatica*). The biostimulant Yoduo was applied to soybean leaves, reflecting 8.61 bags per hectare more than the control. Stimulate® was applied in sugarcane stalks, resulting in higher productivity and higher profitability index compared with treatment without this biostimulant. Biostimulants CROP + ®, SEED + ®, Carbonsolo ®, Kymon Plus ®, which are composed of arginine, serine, phenylalanine, alanine, aspartic acid, glycine, proline and hydroxyproline, glutamic acid, tryptophan, and valine were used in the isolation and in different combinations, applied via leaf and in soybean treatment. These products caused a greater increase in dry mass and leaf area in soybean plants under water stress [24].

Plants subjected to water stress have their cells damaged by free radicals, but the action of antioxidants, reinforced by biostimulants, is able to decrease the toxicity of these radicals, increasing the defense system of plants, due to the increase in their antioxidant levels. According to Ref. [25], plants with high levels of antioxidants improve root and shoot growth, maintaining a high water content in the leaves and low incidence of diseases, both under ideal conditions of cultivation and under environmental stress.

The water deficit affects several aspects of plant growth, with the most apparent effects of water stress being expressed by the reduction of plant size, leaf area, and crop productivity [26]. In recent years, research and use of products considered as plant biostimulants in plants under water stress have been increasing to obtain higher agricultural productivity. For example, the biostimulant Crop + applied by foliar in tomatoes under water stress provided the highest total soluble (°brix)/titratable acidity index, concluding that the application of this biostimulant increases these indices in tomato fruits, even when under water stress [27]. According to [28], the application of the Seed + ® biostimulant via seed treatments and the Crop + ® biostimulant via foliar application on the total chlorophyll index in soybean under water stress increased the total chlorophyll index in soybean plants, providing greater photosynthetic efficiency of plants.

On the other hand, Ref. [29] evaluated the effect of the amino acid L-glutamic acid, via seed treatment, on the germination and development of *Phaseolus vulgaris* seedlings under water restriction. Thus, different concentrations of the amino acid were applied to the seeds placed on polyethylene glycol hydrated filter paper (PEG 6000) under different osmotic potentials (0, 0.2, −0.4, and −0.6 MPa). Thus, the authors concluded that the concentrations of this amino acid did not favor the development of seedlings, interfering negatively in the germination when the osmotic potential was equal to or lower than −0.2 MPa. In addition, seedling development was drastically affected at the osmotic potential equal to or lower than −0.2 MPa, showing a decrease in germination, root length, and seedling volume.

The effects of kinetin and calcium on the physiological characteristics and productivity of soybean plants subjected to water stress and shading in the flowering phase were evaluated [30], and the application of these products promoted the maintenance of the relative water content and the reduction of leakage of cellular electrolytes. In addition, the application of calcium and kinetin to soybean plants under water deficit and shading did not increase the final grain yield.

Maize (*Zea mays*) is a species sensitive to water deficit and among the management techniques related to the induction of tolerance to water deficit in this plant is the application of biostimulants. Thus [31], tried to characterize the effect of two levels of foliar application of the Carbonsolo® biostimulant on the physiological responses of different maize hybrids with and without water deficit. Thirty days after sowing, the Carbonsolo® biostimulant, which contains 25% fulvic acids, 50% humic acids, 20% amino acids, and 2% water-soluble nitrogen was applied to the plant. The authors concluded that the foliar application of this biostimulant, in the initial stage of the maize crop, resulted in a higher relative water content in the leaves and a lower difference between leaf temperature and air temperature under water deficit conditions.

An experiment was conducted with Stimulate® biostimulant and different water regimes (full, partial, and non-irrigated irrigation) to evaluate the action of this biostimulant on leaf water potential, relative water content, liquid photosynthesis, transpiration, stomatal conductance, plant height, main root length, total leaf area, and dry shoot and shoot mass of *Eucalyptus urophylla*. Stimulate® reduced leaf water potential and relative water content; however, it promoted increases in transpiration, stomatal conductance, and liquid photosynthesis in these plants [32]. This effect may have helped to promote greater growth, both in plant height and in length of the main root. Stimulate® promoted a deepening of the roots of the non-irrigated plants, is an important response in a water deficiency situation, since it allows the capture of water in deeper layers of the soil, favoring the maintenance of its growth for a longer time. In addition, the Stimulate® biostimulant was used in order to evaluate the application of biostimulants under initial growth and dry tolerance of sugarcane plants under moderate water stress in an experiment. The maintenance of higher rates of photosynthesis, transpiration, and stomatal conductance was observed [33].

According to Ref. [20], the biostimulants for improving plant resilience in water limiting environments should stimulate root versus shoot growth, which would allow plants to explore deeper soil layer during the drought season and stimulate the synthesis of compatible solutes to re-establish favorable water potential gradients and water uptake at diminishing soil water. Similar positive effects can be given by those microbial biostimulants that create absorption surfaces around the root systems and sequester soil water in favor of the plants.

2.2 Biostimulants and salt stress in plants

Salt stress is one of the most serious limiting factors for crop growth and production. Salts in the soil water may inhibit plant growth by reducing the ability of the plant to take up water and this leads to reductions in the growth rate. Moreover, if excessive amounts of salt enter the plant in the transpiration stream, there will be an injury to cells in the transpiring leaves and this may cause further reductions in growth. These salinity effects cause ion imbalance or disturbances in ion homeostasis and toxicity; this altered water status leads to initial growth reduction and limitation of plant productivity [34]. The management strategies used for cultivation under salinity conditions may increase the productivity and land use both under and under non-saline conditions. Among these strategies, the application of organic matter and biofertilizers, mycorrhization, foliar application of organic and inorganic substances, and the application of biostimulants are highlighted [35].

Biostimulants based on humic substances have been studied in terms of stress protection against salinity due to their biostimulatory activity [36–38]. For salt-affected soil characteristics, results of [39] showed marked improvements in physical and chemical properties of soil by humic substances and *Moringa oleifera* leaf extract is considered as biostimulants that is used for plant growth under normal and salt stress conditions. The application of humic substance-based biostimulants for plants subjected to saline stress showed a capacity to osmotic adjust by maintaining water absorption and cell turgor [40]. Therefore, these authors consider humic substances-based biostimulants as a vigorous growth biostimulant and a nutritive means used to protect various crop plants against some environmental stresses, in special, saline stress.

Application of humic acids to common bean (*Phaseolus vulgaris* L.) under high salinity (120 mM NaCl) increased endogenous proline levels and reduced membrane leakage [38], which are both indicators of better adaptation to saline environments. Humic acid extracts applied to rice (*Oryza sativa* L.) played a role in activating anti-oxidative enzymatic function and increased reactive oxygen species (ROS) scavenging enzymes. These enzymes are required to inactivate toxic free oxygen radicals produced in plants under drought and saline stress [41].

The commercial biostimulant Stimulate® presents 0.009% cytokinin, 0.005% gibberellin, and 0.005% auxin, and it has been used in several studies regarding saline stress in plants [42–47]. However, the results are not conclusive about its effect on improving plant resistance under salt stress. On the other hand, the application of the commercial biostimulant Retrosal®, containing calcium, zinc, and specific active ingredients, on lettuce conferred enhanced tolerance when plants were exposed to NaCl treatments, due to its multifaceted action at both biochemical and physiological level. In particular, a significant biostimulant effect was observed on several variables examined, among which fresh yield, dry biomass, chlorophyll content *in vivo*, nitrate concentration, and some leaf gas exchange parameters as well as chlorophyll *a* fluorescence parameters [48].

In addition to these substances mentioned above, biostimulants presenting algae and arbuscular mycorrhizal fungi (AMF), fungi, and bacteria as raw material are bioactive compounds in improving salinity stress tolerance by increasing germination rate, growth characters (length, fresh, and dry weight) of shoots and roots, plant quality, productivity, and yield [2, 20]. Algal extracts target a number of pathways to increase tolerance under stress [21]. Application of algal extracts significantly increased the contents of total chlorophyll and antioxidant phenomenon in wheat plants irrigated with brackish water, exhibiting a strong positive correlation with the increase in fresh weight, grain weight, and yield components [49]. Algal extracts have been used on Kentucky bluegrass (*Poapratensis* L. cv. Plush) to alleviate salinity stress from saline watering in turfgrass experiments [50].

Many studies have shown that the application of commercial biostimulants based on arbuscular mycorrhizal fungi (AMF) inoculum benefits crops under agricultural saline stress conditions by supporting plant nutrition, influencing plant development (bioregulators), and inducing tolerance to saline stresses (bioprotector) [51]. AMF can contribute to protect tomato plants against salinity by alleviating the salt-induced oxidative stress [52]. According to these authors, this ameliorative effect of mycorrhizal colonization shows significant interactions with cultivar and salt exposure. Enhanced antioxidant enzymes activity and lower lipid peroxidation in mycorrhizal plants may contribute to better maintenance of the ion balance the photochemical reactions in leaves under salinity. Plant growth-promoting rhizobacteria-based biostimulants are considered easy-to-use agroecological tools for stimulating plant growth and enhancing plant nutrient uptake and salt stress tolerance [53]. Salt-tolerant plant growth-promoting rhizobacteria significantly influenced the growth and yield of wheat crops in saline soil [54].

Under salt stress, many authors classified the effects of different categories of biostimulants on plants into direct and indirect influences. The indirect impacts are linked to improvements of physical, chemical, and biological properties of soils, while the direct influences are attributed to improvements of germination, plant growth (root and shoot) as an improvement on resistance of plants to salt stress, as previously mentioned [35].

As one can see, many authors consider biostimulant application as a sustainable tool for plant production and a meaningful approach to counteract salt stress in plants. In this sense, biostimulant application in agriculture under saline conditions has demonstrated the potential of various categories of biostimulants to improve crop production and to ameliorate salinity stress.

2.3 Biostimulants and temperature stress in plants

Temperature stress in plants is classified into three types depending on the stressor, which may be high, chilling, or freezing temperature. Temperature-stressed plants show low germination rates, growth retardation, reduced photosynthesis, and often die. The development of temperature stress can be induced by a high- or low-temperature, and may depend on the duration of the exposure, the rate of temperature changes, and the plant growth stage at which stress exposure occurs. However, plants possess a variety of molecular mechanisms involving proteins, antioxidants, metabolites, regulatory factors, other protectants, and membrane lipids to cope with temperature stress [55].

The temperature factor can be a relevant obstacle to the germination and early development of many horticultural species. Studies have shown deleterious effects on germination when seeds of various crops are exposed to high temperature. Biostimulants are therefore options for mitigating such effects and, by presenting defensive properties against abiotic stresses, such as drought, salinity, and high variation of temperatures; they can alleviate plant defense system of such stressors [1].

Increasing doses of Stimulate® biostimulant (0, 4, 8, and 12 mL L^{-1}) as a thermal stress reliever (temperatures 25 and 40°C) on germination and initial growth of melon favored the germination rate by the increase of the doses of biostimulant at both temperatures [56]. Thus, the biostimulant can be used to improve the germination of the melon in high temperature conditions and to improve the initial development of the melon in regions that present high temperatures.

A research was conducted to determine the effects of two biostimulants (humic acid and biozyme) or three different salt (NaCl) concentrations on parsley, leek, celery, tomato, onion, lettuce, basil, radish, and garden cress seed germination at 10, 15, 20, and 25°C. It resulted that two applications of both biostimulants increased seed germination of parsley, celery, and leek at all temperature treatments. In addition, interaction among biostimulants and temperatures was significant in all of the vegetable species [57].

The effectiveness of a product obtained from the enzymatic hydrolysis of porcine hemoglobin (PHH) as a biostimulant that lessen the effect of thermal stress in plants, was observed by two experiments carried out in which lettuce plants were subjected to short-term episodes of intense cold and heat, with different doses of PHH. The results showed that at the highest tested doses, the PHH product ameliorated the negative effects on lettuce growth caused by the increase in temperature and lessened the harmful effects of the cold, i.e., promoted a reaction that lessened the harmful effects caused by the intense cold and heat treatments [58].

In the same way, Ref. [59] evaluating PHH, specifically porcine blood, on strawberry plants in the initial growing stages after being transplanted and subject to conditions of intense cold, an experiment was carried out to compare two doses of PHH with a commercial biostimulant (CB) and a control treatment (C). The results showed that the highest dose of PHH produced more biomass of newly formed roots, that both doses of PHH produced early flowering, and that both doses of PHH led to a significant increase in the early production of fruit compared with the C treatment. None of the biostimulant treatments improved the survival ratio of the strawberry plants compared with the control treatment.

According to Ref. [60], plant thermal acclimation mechanisms include the accumulation of compatible N-rich solutes, such as amino acids, that confer stress tolerance.

Thus, in order to assess the effect of exogenous amino acids treatments, several experiments with plants (lettuce and ryegrass), subjected to three different types of cold stress, were conducted applying an amino acid product obtained by Enzymatic Hydrolysis (Terra-Sorb® Foliar). Results showed that treated lettuce plants have a higher fresh weight than control plants, exhibiting a higher stomatal conductance, which implies productive improvements. In addition, at a high temperature (36°C), ryegrass treated with Terra-Sorb® Foliar showed a superior photosynthetic efficiency (Fv/Fm) and maintains higher levels of chlorophylls and carotenoids. These findings suggest that Terra-Sorb® Foliar has a similar effect to natural plant amino acids and promotes a better more prompt crop recovery from temperature stress.

A major concern in turfgrass management is the summer decline in turf quality and growth of cool-season grass species [61]. Based on this, these researchers investigated whether foliar application of trinexapac-ethyl (TE) and two biostimulants (TurfVigor and CPR) containing seaweed extracts would alleviate the decline in creeping bentgrass (*Agrostis stolonifera* L.) growth during summer months and examined effects of TE and the biostimulants on leaf senescence and root growth. Foliar application of TE resulted in significant improvement in turf quality, density, and chlorophyll content compared with the control. Both TurfVigor and CPR significantly improved visual quality by promoting both shoot and root growth. This study suggests that the proper application of TE and selected biostimulants could be effective to improve the summer performance of creeping bentgrass.

Perennial ryegrass plants treated with a product-based protein and exposed to prolonged high air temperature stress exhibited both an improved photochemical efficiency and membrane thermostability than untreated plants [62]. These results provided consistent and interesting results and showed that foliar applications of protein hydrolysates can positively affect plant tolerance to heat stress [63].

The stress protection of bacterial biostimulants to rainfed field crops can be of particular relevance under increasing temperatures foreseen by most prediction models of climate change. Wheat inoculated with the thermotolerant *Pseudomonas putida* strain AKMP7 significantly increased heat tolerance. Inoculated plants had increased biomass, shoot and root length, and seed size [64].

Bioactive compounds present in the seaweed extracts enhance the performance of plants under abiotic stresses. Spray applications of extracts have been shown to improve plant tolerance to freezing temperature stress. Moreover, commercial *A. nodosum* extract was also reported to promote the performance of lettuce seedling under high temperature stress. In addition, seed germination of lettuce was influenced by priming with *A. nodosum* extract in that germination improved under high temperature conditions [65].

3. Final remarks

Biotic stress such as, drought, high soil salinity, heat, and cold is the common adverse environmental conditions that affect and limit crop productivity worldwide.

Plant biostimulants include diverse substances and microorganisms that enhance plant growth and resistance to abiotic stresses and increase seed and/or grain yield and quality. The definition and concept of plant biostimulants are still evolving, which is partly a reflection of the diversity of inputs that can be considered biostimulants.

Agricultural biostimulants may contribute to make agriculture more sustainable and resilient, since a brief review of the literature shows a clear role for a diverse number of biostimulants that have protective effects against abiotic stress.

Biostimulant treatments of agricultural crops have the potential to improve plant resilience to environmental perturbations. In order to fine-tune application rates, biostimulant-plant specificities and techniques are identified that may yield the highest impact on stress protection; high priority should be given to better understanding of the causal/functional mechanism of biostimulants.

Although input-producing companies are investing in the development of new products for the incorporation of biostimulants and additives to agriculture each year, it can be observed from studies carried out that little is known about the mechanisms of action of these inputs in order to optimize the real gains from the incorporation of these products into agricultural production.

In addition, there is a need to address the underlying mechanisms responsible for these effects, given the large number of substances that can be used as biostimulant raw material, such as humic substances, seaweed extracts, plant hormones, and plant growth-promoting rhizobacteria.

The application of an appropriate biostimulant can improve root and shoot vigor, however, the selection of the appropriate biostimulant is critical as the effects can vary markedly between species.

Author details

Ana Carolina Feitosa de Vasconcelos* and Lúcia Helena Garófalo Chaves
Federal University of Campina Grande, Campina Grande, Brazil

*Address all correspondence to: ana3carol@yahoo.com.br

References

[1] Du Jardin P. Plant biostimulants: Definition, concept, main categories and regulation. Scientia Horticulturae. 2015;**196**:3-14. DOI: 10.1016/j.scienta.2015.09.021

[2] Yakhin OI, Lubyanov AA, Yakhin IA, Brown PH. Biostimulants in plant science: A global perspective. Frontiers in Plant Science. 2017;7(2049):1-32. DOI: 10.3389/fpls.2016.02049

[3] Basak A. Biostimulators – definitions, classification and legislation. In: Gawronska H, editor. Monographs Series: Biostimulators in Modern Agriculture. General Aspects. Warsaw: Wieś Jutra; 2008. pp. 7-17

[4] Bulgari R, Cocetta G, Trivellini A, Vernieri P, Ferrante A. Biostimulants and crop responses: A review. Biological Agriculture and Horticulture. 2015;**31**:1-17. DOI: 10.1080/01448765.2014.964649

[5] Yaronskaya E, Vershilovskaya I, Poers Y, Alawady AE, Averina N, Grimm B. Cytokinin effects on tetrapyr role biosynthesis and photosynthetic activity in barley seedlings. Planta. 2006;**224**:700-709

[6] Singh J, Takhur JK. Photosynthesis and abiotic stress in plants. In: Vats S, editor. Biotic and Abiotic Stress Tolerance in Plants. Singapore: Springer Nature Singapore Private Ltd; 2018. pp. 27-46

[7] Du Jardin P. The science of plant biostimulants—A bibliographic analysis, Ad hoc Study Report. 2012. Brussels: European Commission. Available from: http://hdl.handle.net/2268/169257 [Accessed: 25-04-2019]

[8] Couto CA, Peixoto CP, Vieira EL, Carvalho EV, Peixoto VAB. Ação da cinetina, ácido indolbutírico e ácido giberélico na emergência do girassol sob estresse por alumínio. Comunicata Scientiae. 2012;**3**:206-209

[9] Cramer GR, Urano K, Delrot S, Pezzotti M, Shinozaki K. Effects of abiotic stress on plants: A systems biology perspective. BMC Plant Biology. 2011;**11**:163. DOI: 10.1186/1471-2229-11-163

[10] Huang J, Levine A, Wang Z. Plant abiotic stress. The Scientific World Journal. 2013;**2013**:432836. DOI: 10.1155/2013/432836

[11] Teale WD, Paponov IA, Palme K. Auxin in action: Signaling, transport and the control of plant growth and development. Nature Reviews. Molecular Cell Biology. 2006;**7**:847-859

[12] Taiz L, Zeiger E. Fisiologia vegetal. 4th ed. Porto Alegre: Artmed; 2009. p. 819

[13] Santos CMG, Vieira EL. Efeito de bioestimulante na germinação de grãos, vigor de plântulas e crescimento inicial do algodoeiro. Magistra. 2005;**17**:124-130

[14] Rathore SS, Chaudhary DR, Boricha GN, Ghosh A, Bhatt BP, Zodape ST, et al. Effect of seaweed extract on the growth, yield and nutrient uptake of soybean (*Glycine max*) under rainfed conditions. South African Journal of Botany. 2009;**75**:351-355

[15] Amin AA, Fatma AE, Gharib M, El-Awad A, El-Sherbeny M, Rashad A. Physiological response of onion plants to foliar application of putrescine and glutamine. Scientia Horticulturae. 2011;**129**:353-360

[16] Forde BG, Lea PJ. Glutamate in plants: Metabolism, regulation, and signaling. Journal of Experimental Botany. 2007;**58**:2339-2358

[17] Robinson SA, Slade AP, Fox GG, Phillips R, Ratcliffe RG, Stewart GR.

The role of glutamate dehydrogenase in plant nitrogen metabolism. Plant & Cell Physiology. 1991;**95**:509-516

[18] Rhods D, Handa S, Bressan RA. Metabolic changes associated with adaptation of plant cells to water stress. Plant Physiology. 1986;**82**:890-903

[19] Lea PJ, Sodek L, Parry MAJ, Shewry PR, Halford NG. Asparagine in plants. The Annals of Applied Biology. 2007;**150**:1-26

[20] Van Oosten MA, Pepe O, Pascale SD, Silletti S, Maggio A. The role of biostimulants and bioeffectors as alleviators of abiotic stress in crop plants. Chemical and Biological Technologies in Agriculture. 2017;**4**:5. DOI: 10.1186/s40538-017-0089-5

[21] Salehi-Lisar SY, Bakhshayeshan-Agdam H. Drought stress in plants: Causes, consequences, and tolerance. In: Hossain MA, Wani SH, Bhattacharjee S, Burritt DJ, Tran LP, editors. Drought Stress Tolerance in Plants: Physiology and Biochemistry. Switzerland: Springer International Publishing; 2016. pp. 1-16. DOI: 10.1007/978-3-319-28899-4

[22] Lana AMQ. Aplicação de reguladores de crescimento na cultura do feijoeiro. Bioscience Journal. 2009;**25**(1):13-20

[23] Russo RO, Berlyn GP. The use of organic biostimulants to help low-input sustainable agriculture. Journal of Sustainable Agriculture. 1990;**1**(2):19-42

[24] Santos VM, Vaz De Melo A, Cardoso DP, Silva AR, Benício LPF, Ferreira EA. Desenvolvimento de plantas de soja em função de bioestimulantes em condições de adubação fosfatada. Bioscience Journal. 2014;**30**:1087-1094

[25] Hamza B, Suggars A. Biostimulants: Myths and realities. 2001. Available from: http://archive.lib.msu.edu/tic/tgtre/article/2001aug6.pdf [Accessed: 29 July 2013]

[26] Kramer P. Water Relations of Plants. New York: Academic Press; 1983

[27] Peripolli M, Paranhos JT, Dornelles SHB, Morais TB, Muller E, Spatt LL, et al. Influência do uso do bioestimulante crop+® na qualidade de frutos de tomateiro sob estresse hídrico. In: Anais do II Simpósio latino-americano sobre bioestimulantes na agricultura & IX Reunião brasileira sobre indução de resistência em plantas a patógenos. Florianópolis; 2018. p. 166

[28] Morais TB, Swarowsky A, Sanchotene D, Peripolli M, Muller E, Shimoia E, et al. Efeito dos bioestimulantes seed+® e crop+® no índice de clorofila total da soja sob estresse hídrico. In: Anais do II Simpósio latino-americano sobre bioestimulantes na agricultura & IX Reunião brasileira sobre indução de resistência em plantas a patógenos. Florianópolis; 2018. p. 176

[29] Carvalho TC, Silva SS, Silva RC, Panobianco M, Mógor AF. Influência de bioestimulantes na germinação e desenvolvimento de plântulas de *Phaseolus vulgaris* sob restrição hídrica. Revista de Ciências Agrárias. 2013;**36**:199-205

[30] Fioreze SL, Rodrigues JD, Carneiro JPC, Silva AA, Lima MB. Fisiologia e produção da soja tratada com cinetina e cálcio sob déficit hídrico e sombreamento. Pesquisa Agropecuária Brasileira. 2013;**48**:1432-1439

[31] Almeida MS, Nunes AS, Casagrande RR. Aplicação foliar de bioestimulante em híbrido de milho com e sem déficit hídrico. In: XIV Seminário Nacional Milho Safrinha. Cuiabá; 2017. pp. 146-151

[32] Santos RKA. Bioestimulante vegetal na produção de mudas de Eucalyptus

urophyllae no seu crescimento inicial em diferentes regimes hídricos [dissertação]. Vitória da Conquista: Universidade Estadual do Sudoeste da Bahia; 2015

[33] Wanderley Filho HCL. Uso de bioestimulantes e enraizadores no crescimento inicial e tolerância à seca em cana-de-açúcar [dissertação]. Rio Largo: Universidade Federal de Alagoas; 2011

[34] Parihar P, Singh S, Singh R, Singh VP, Prasad SM. Effect of salinity stress on plants and its tolerance strategies: A review. Effect of salinity stress on plants and its tolerance strategies: A review. Environmental Science and Pollution Research. 2015;**22**:4056-4075. DOI: 10.1007/s11356-014-3739-1

[35] Lacerda CF, Costa RNT, Bezerra MA, Gheyi HR. Estratégias de manejo para uso de água salina na agricultura. In: Gheyi HR, Dias NS, Lacerda CF, editors. Manejo da salinidade na agricultura: Estudos básicos e aplicados. Fortaleza: INCT Sal; 2010. pp. 306-318

[36] Türkmen Ö, Dursun A, Turan M, Erdinc C. Calcium and humic acid affect seed germination, growth and nutrient content of tomato (*Lycopersicon esculentum* L.) seedlings under saline soil conditions. Acta Agriculturae Scandinavica Section B Soil and Plant Science. 2004;**54**:168-174. DOI: 10.1080/09064710310022014

[37] Paksoy M, Türkmen Ö, Dursun A. Effects of potassium and humic acid on emergence, growth and nutrient contents of okra (*Abelmoschus esculentus* L.) seedling under saline soil conditions. African Journal of Biotechnology. 2010;**9**:5343-5346

[38] Aydin A, Kant C, Turan M. Humic acid application alleviate salinity stress of bean (*Phaseolus vulgaris* L.) plants decreasing membrane leakage. African Journal of Agricultural Research. 2012;7:1073-1086. DOI: 10.5897/ajar10.274

[39] Desoky ES, Merwad AM, Elrys AS. Response of pea plants to natural bio-stimulants under soil salinity stress. American Journal of Plant Physiology. 2017;**12**:28-37

[40] Azevedo RA, Lea PJ. Research on abiotic and biotic stress—What next? The Annals of Applied Biology. 2011;**159**:317-319

[41] García AC, Santos LA, Izquierdo FG, Sperandio MVL, Castro RN, Berbara RLL. Vermicompost humic acids as an ecological pathway to protect rice plant against oxidative stress. Ecological Engineering. 2012;**47**:203-208. DOI: 10.1016/j.ecoleng.2012.06.011

[42] Souza Neta ML, Oliveira FA, Torres SB, Souza AAT, Silva DDA, Santos ST. Gherkin cultivation in saline medium using seeds treated with a biostimulant. Acta Scientiarum Agronomy. 2018;**40**:e35216. DOI: 10.4025/actasciagron.v40i1.35216

[43] Oliveira FA, Medeiros JF, Cunha RC, Souza MWL, Lima LA. Uso de bioestimulante como agente amenizador do estresse salino na cultura do milho pipoca. Revista Ciência Agronômica. 2016;**47**:307-315. DOI: 10.5935/1806-6690.20160036

[44] Oliveira FA, Medeiros JF, Alves RC, Lima LA, Santos ST, Régis LRL. Produção de feijão caupi em função da salinidade e regulador de crescimento. Revista Brasileira de Engenharia Agrícola e Ambiental. 2015a;**19**(11):1049-1056. DOI: 10.1590/1807-1929/agriambi.v19n11p1049-1056

[45] Oliveira FA, Guedes RAA, Gomes LP, Bezerra FMS, Lima LA,

Oliveira MKT. Interação entre salinidade e bioestimulante no crescimento inicial de pinhão-manso. Revista Brasileira de Engenharia Agrícola e Ambiental. 2015b;**19**(3):204-210. DOI: 10.1590/1807-1929/agriambi.v19n3p204-210

[46] Oliveira FA, Medeiros JF, Oliveira MKT, Souza AAT, Ferreira JÁ, Souza MS. Interação entre salinidade e bioestimulante na cultura do feijão caupi. Revista Brasileira de Engenharia Agrícola e Ambiental. 2013;**17**:465-471. DOI: 10.1590/S1415-43662013000500001

[47] Klahold CA, Guimarães VF, Echer MM, Klahold A, Contiero RL, Becker A. Resposta da soja (*Glycine max* (L.) Merrill) à ação de bioestimulante. Acta Scientiarum Agronomy. 2006;**28**:179-185. DOI: 10.4025/actasciagron.v28i2.1032

[48] Bulgari R, Trivellini A, Ferrante A. Effects of two doses of organic extract-based biostimulant on greenhouse lettuce grown under increasing NaCl concentrations. Frontiers in Plant Science. 2019;**9**:1870. DOI: 10.3389/fpls.2018.01870

[49] El-Baky HHA, Hussein MM, El-Baroty GS. Algal extracts improve antioxidant defense abilities and salt tolerance of wheat plant irrigated with sea water. African Journal of Biochemistry Research. 2008;**2**(7):151-164

[50] Nabati DA, Schmidt RE, Parrish DJ. Alleviation of salinity stress in Kentucky bluegrass by plant growth regulators and iron. Crop Science. 1994;**34**(1):198-202. DOI: 10.2135/cropsci1994.0011183X003400010035x

[51] Rouphael Y, Franken P, Schneider C, Schwarz D, Giovannetti M, Agnolucci M. Arbuscular mycorrhizal fungi act as biostimulants in horticultural crops. Scientia Horticulturae. 2015;**196**:91-108. DOI: 10.1016/j.scienta.2015.09.002

[52] Latef AA, Chaoxing H. Effect of arbuscular mycorrhizal fungi on growth, mineral nutrition, antioxidant enzymes activity and fruit yield of tomato grown under salinity stress. Scientia Horticulturae. 2011;**127**:228-233. DOI: 10.1016/j.scienta.2010.09.020

[53] Mire GL, Nguyen ML, Fassotte B, Du Jardin P, Verhegeen F, Delaplace P, et al. Implementing plant biostimulants and biocontrol strategies in the agroecological management of cultivated ecosystems. A review. Biotechnology, Agronomy, Society and Environment. 2016;**20**(S1):299-313

[54] Upadhyay SK, Singh DP. Effect of salt-tolerant plant growth-promoting rhizobacteria on wheat plants and soil health in a saline environment. Plant Biology. 2015;**17**(1):288-293. DOI: 10.1111/plb.12173

[55] Kai H, Koh I. Temperature stress in plants. In: eLS. Chichester: John Wiley & Sons, Ltd; 2014. DOI: 10.1002/9780470015902.a0001320.pub2

[56] Vendruscolo EP, Martins APB, Campos LFC, Seleguini A, Santos MM. Amenização de estresse térmico via aplicação de bioestimulante em sementes de meloeiro Cantaloupe. Brazilian Journal of Biosystems Engineering. 2016;**10**:241-247

[57] Yildirim E, Dursun A, Güvenc I, Kumlay AM. The effects of different salt, biostimulant and temperature levels on seed germination of some vegetable species. Acta Horticulturae. 2002;**579**:249-253

[58] Polo J, Barroso R, Azcón-Bieto J, Ródenas J, Cáceres R, Marfà O. Porcine hemoglobin hydrolysate as a biostimulant for lettuce plants subjected

to conditions of thermal stress. HortTechnology. 2006;**16**:483-487

[59] Marfà O, Cáceres R, Polo J, Ródena J. Animal protein hydrolysate as a biostimulant for transplanted strawberry plants subjected to cold stress. Acta Horticulturae. 2009;**842**:315-318

[60] Botta A. Enhancing plant tolerance to temperature stress with amino acids: An approach to their mode of action. Acta Horticulturae. 2013;**1009**:29-35

[61] Xu Y, Huang B. Responses of creeping Bentgrass to Trinexapac-ethyl and biostimulants under summer stress. HortScience. 2010;**45**:125-131

[62] Kramer PJ. Drought, stress, and the origin of adaptations. In: Turner NC, Kramer PJ, editors. Adaptation of Plants to Water and High Temperature Stress. New York, NY, USA: John Wiley; 1980. pp. 7-20

[63] Kauffman GL III, Kneivel DP, Watschke TL. Effects of a biostimulant on the heat tolerance associated with photosynthetic capacity, membrane thermostability, and polyphenol production of perennial ryegrass. Crop Science. 2007;**47**:261-267

[64] Shaik ZA, Sandhya V, Grover M, Linga VR, Bandi V. Effect of inoculation with a thermotolerant plant growth promoting pseudomonas putidastrain AKMP7 on growth of wheat (*Triticum* spp.) under heat stress. Journal of Plant Interactions. 2011;**6**:239-246

[65] Battacharyya D, Babgohari MZ, Rathor P, Prithiviraj B. Seaweed extracts as biostimulants in horticulture. Scientia Horticulturae. 2015;**196**:39-48

Chapter 2

Application of Bacteria as a Prominent Source of Biofertilizers

Prabakaran Elavarasi, Muthuraman Yuvaraj and Pandurangan Gayathri

Abstract

There are different types of microorganisms are used in the biofertilizers. Biofertilizers being essential components of organic farming play vital role in maintaining long-term soil fertility and sustainability; biofertilizers would be the viable option for farmers to increase productivity per unit area. These potential biological fertilizers would play a key role in productivity and sustainability of soil and also in protecting the environment as eco-friendly and cost-effective inputs for the farmers. At the same time, overlooking the significance of ensuring and maintaining a high quality standard of the product will have negative impact. Hence, a proper knowledge of bio-inoculants and its functioning will pave way to tape the resources in a better way. Thus, the chapter provides overview knowledge about different bacterial biofertilizers, its associations with plants and transformations of nutrients in soil. Adopting a rational approach to use and management of microbial fertilizers in sustainable agriculture thrives vast potential for the future.

Keywords: biofertilizers, microorganisms, rhizobium, mycorrhiza, biological nitrogen fixation

1. Introduction

One of the present day challenges in agriculture is eco-friendly practices. Though the benefits of Green revolution have been reaped by us in terms of production, the other side of it i.e., over usage of chemical fertilizers and its subsequent deterioration of soil health has been realized these days [1]. Hence, awareness of practicing organic agriculture has been taken to various spheres and products of organic agriculture are fetching up huge market. One of the organic agriculture practices includes usage of biofertilizers in farming [2]. The biofertilizers has several other advantages as well like they are cost effective, eco-friendly and renewable source of plant nutrients hence forms one of the important components of integrated nutrient management. As of now we could not claim bio-inoculants as a right alternative to chemical fertilizers but in near future the scientific understanding of the same will pave way for its right use and reap full benefits [3]. In addition to this in global scale, recent published works on bio fertilizers states about the varied role of bio-inoculants *viz.*, other than nutrient transformations in different crops. To mention few, increase in root growth has been observed in wheat due to inoculation of bio-inoculant consortia. The *Rhizobium* inoculation increases deaminase activity in pulses crops. Hence this chapter focuses on different bio-inoculants and its uses in farming.

IntechOpen

2. Importance of soil microbes in nutrient transformations

It is well established fact that soil microbes have versatile enzyme systems hence perform various nutrient transformations in soil which is very important for maintaining soil equilibrium and its health. Among the nutrient transformations, nitrogen and phosphors transformations forms significant importance, since they are the major plant nutrients derived from the soil.

3. Biological nitrogen fixation

Biological nitrogen fixation is a component of nitrogen cycle which involves fixing up of atmospheric nitrogen by particular soil microorganisms. Nitrogen fixing ability has been restricted only to certain bacteria and few actinomycetes which belong to various groups and they are referred to as diazotrophs. The diazotrophic microbes are ubiquitous to soil and are classified according to mode of nitrogen fixation to plants.

The process of biological nitrogen fixation has been first documented in anaerobic bacterium *Clostridium pasteurianum* from which the enzyme nitrogenase has been isolated. However, today, the organism has not been commercially used for the purpose. The nitrogen fixation is mediated by nitrogenase enzyme which reduces gaseous nitrogen to ammonia. All diazotrophs seemed to possess the enzyme and found to deliver quite similar mechanism of nitrogen fixation.

4. Phosphorus solubilizing bio-inoculants

The fate of phosphorus is that it forms apatites with the salts present in the soil. In acid soil phosphorus will becomes Aluminium phosphates and Iron phosphates while in alkaline soils it becomes calcium phosphates or sodium phosphates and becomes unavailable to plants. In order to make these form of phosphorus to available form some of the bio-inoculants produces organic acids which convert them to soluble form like hypophosphites which can be taken by plants. Examples of phosphorus solubilising bacteria: *Bacillus megatherium var phosphaticum*, *Bacillus megaterium* var. *phosphaticum*, *Bacillus subtilis*, *Bacillus circulans*, *Pseudomonas striata*.

5. Phosphorus mobilising bacteria

The soil microorganisms able to solubilize precipitated forms of P or mineralize organic P has been characterized. The most important phosphorus mobilising bacteria is *Pseudomonas* and *Bacillus* being predominant. These organisms ordinarily related to the rhizosphere and, when inoculated onto plants, often result in improved growth and P nutrition with responses being observed under both glasshouse and field conditions. Despite this, there are few examples of successful application of microbial inoculants. Essentially, a lack of consistent performance under different environmental conditions in the field has precluded their wider use. A number of things may be known to clarify this variable performance [4].

The *Bacillus spp*. convert the complex form of essential nutrients, such as P and N, to a simple available form that is used during uptake by plant roots [5, 6]. Phosphate is involved in nucleic acid, phospholipid, and adenosine triphosphate (ATP) metabolism, among other metabolic pathways, in plant cells. The secretion

Bacillus species	Plant growth promotion	References
B. insolitus, *B. subtilis*, *B. methylotrophicus*	Increase the length and biomass of shoot, root and leaves	[18]
B. megaterium, *B. subtilis*	Enhance fruit and grains yield	[6]
B. pumilus, *B. megaterium*	Solubilize the P and fix the N in soil and increase their transport to roots	[20]

Table 1.
Biofertilizer effect of Bacillus spp *on crop plants.*

of phosphatases and organic acids from *Bacillus spp.* acidifies the surrounding environment to facilitate the conversion of inorganic phosphate into free phosphate [7, 8]. Additionally, N is an important component of proteins, nucleic acids and other organic compounds in plants, and the available form of N in soil is limited, which slows plant growth in natural habitats. Some of the Bacillus spp. release ammonia from nitrogenous organic matter [9]. The *Bacillus spp*. have the nif H gene and produce nitrogenase (EC 1.18.6.1), which can fix atmospheric N_2 and provide it to plants to enhance plant growth and yield by delaying senescence [10]. The iron-chelating properties of Bacillus spp. via siderophore production help to solubilize iron from minerals and organic compounds in rhizospheres [11]. Siderophores bind Fe^{3+} in complex substances and reduce the Fe^{3+} to Fe^{2+} which then enters plants (**Table 1**).

Seed germination is regulated by sugars, nitrate, and phytohormones, such as auxin, cytokinins, ethylene, abscisic acid (ABA), GAs, brassinosteroid (BR) and light [12, 13]. Salt deposition in soil decreases the osmotic potential of the growth medium for plants and reduces the water availability [14]. Plants respond to salt-induced osmotic stress by closing their stomata, thus limiting the loss of cellular water content and gas exchange, which reduces the photosynthetic rate.

6. Plant growth promoting rhizobacteria (PGPR)

Plant growth promoting rhizobacteria (PGPR) a heterogeneous group of microorganisms are known to improve plant growth by their ability to colonize the rhizosphere besides their effect as biocontrol agents and producers of plant hormones. PGPR also alter the plant physiological processes resulting in enhanced nutrient uptake. These organisms possess the ability to produce siderosphere, a class of high affinity iron transport molecules which also act as a growth promoting factor. The ability of PGPR to produce siderophore or to affect the activity of siderophores produced by under Fe-deficient condition siderophore producing pseudomonads form a yellow, green fluorescent siderophore iron complex. A few microorganisms are known to chelate iron through production of siderophores. The well-known PGPR include *Azotobacter, Azospirillum, Azoarcus, Klebsiella, Bacillus, Pseudomonas, Arthrobacter, Enterobacter, Burkholderia, Serratia*, and *Rhizobium* [15].

7. Important diazotrophs in commercial use

Rhizobium is the most studied bio-inoculant which forms symbiotic association with legume plants. It was first shown by Boussingault that leguminous plant can fix atmosphere N2 which hellriegel and wilfarth clarified that the process is done by bacteria residing in the roots of leguminous plants. The purified bacterium was put into various examinations and now well-developed nitrogen fixing strains are available in various commercial production units.

This bio-inoculant is specific for legume crops and forms nodules in the roots of the plants. It enriches the soil fertility also after harvesting of the crop. Hence it is the most preferred bioincoulant. Other than root nodulating *Rhizobium* some of the strains found to nodulate stem known as *Azorhizobium* present in *Sesbania rostrata.* Rhizobium species are specific to legume crops because of nod factors they produce. However, some leguminous plants found to develop effective nodules on inoculation with the *Rhizobia* obtained from the nodules from other legume groups, which are referred to as cross inoculation grouping [16].

8. *Azospirillum*

Azospirillum is considered as very important diazotrophs as it form associative symbiotic relationship with the roots of graminaceous plants. It is generally recommended for rice crop. The organism is microaerophillic, some are aerobic motile and gram negative in nature hence suits well for rice field conditions. It was first isolated by Beijernick and was named as *Spririllum lipoferum* later named as *Azospirillum.* In addition to nitrogen fixing ability, they also produce growth promoting substances such as indole acetic acid [17]. Some of the important species of *Azospirillum* has been listed below:

- *A. brasilense*
- *A. lipoferum*
- *A. amazonense*
- *A. halopraeferens*
- *A. irkense*
- *A. dobereinerae*
- *A. largimobilis*

9. *Azotobacter*

Azotobacter are gram-negative free living bacterium in the rhizosphere soil of many plant species, discovered by Beijernick. The bacterium is very well recognised diazotroph and fixes atmospheric nitrogen in its habitat. Owing to its versatile adaptability and nitrogen fixing ability, they are commercially used in agriculture for many crops and are known with a brand name azotobacterin. Some species of *Azotobacter* known to produce alginic acid, a compound used in medical industry and in food industry it is used as additive in ice creams and cakes. Apart from its nitrogen fixing ability, it also synthesise many phytohormones such as auxins and helps in promoting growth of the plants. They are involved in mobilising heavy metals in the soil thus used for bioremediation purposes as well. Many species of *Azotobacter* are pigment producers and found to degrade aromatic compounds in the agriculture lands [18].

10. Gluconoacterobacter diazotrophics

They are endotrophic bacterium which resides insides the stem of sugarcane as it prefers high sucrose and acid content for its survival. They have the ability of capturing atmospheric nitrogen and converting into ammonical form [19]. Moreover they are known for stimulating plant growth by tolerant to acetic acid. The bacterium was first discovered in Brazil by scientists Vladimir *A. cavalcante* and Johanna Dobereiner. They are originally known as *Acetobacter* belong to Acetobacteriaceae family and got the current name due to carbon source requirement. Besides nitrogen fixing ability they are known to synthesise indole-3 acetic acid which promote the growth of the associated plant species [20]. Also reports suggest this bacterium controls pathogen especially *Xanthomonas albilineans* in sugarcane. Thus in recent years it is the most recommended bio-inoculant for sugarcane [16].

11. Algal biofertilizers

The potentiality of algal biofertilizers are realised long before by 1939, when WHO attributed the tropical rice natural fertility to green blue cholorphytic algae. Among algae, only blue green algae have biological nitrogen fixing ability due to the presence of heterocysts cells in them [17]. This bio-inoculant is recommended only for rice crop and was proved to improve soil fertility by nitrogen fixation and organic matter enrichment after harvest. In some places, practice of culturing algae as dual crop along with rice has been done which found to inhibit small weed growth during cropping. Apart from this some of the algal species also promote growth by producing growth promoting substances [18].

The following list is some of the nitrogen fixing algal species: (a) Examples of unicellular nitrogen fixing algae: *Gloeothece*, *Gloeobacter*, *Synechococcus*, *Cyanothece*, *Gloeocapsa*, *Synechocystis*, *Chamaesiphon*, *Merismopedia*; (b) filamentous non heterocystous forms of *Cyanobacteria*, *Oscillatoria*, *Spirulina*, *Arthrospira*, *Lyngbya*, *Microcoleus*, *Pseudanabaena*; (c) Filamentous heterocystous forms *Anabaena*, *Nostoc*, *Calothrix*, *Nodularia*, *Cylinodrosperum*, *Scytonema.*

12. *Anabaena azollae*

Anabaena is a special type of algae which forms symbiotic association with free floating water fern *Azolla*. Water fern is bilobed in nature and algae resides in the roots of the fern. The common species of algae forming symbiotic association with Azolla are *A. microphylla*, *A. filiculoides*, *A. pinnata*, *A. caroliniana*, *A. nilotica*, *A. rubra and A. mexicana*. This alga takes shelter and carbon from the water fern and in turn fixes atmospheric nitrogen. They need sunlight and water for its multiplication and hence can be used for rice crop as dual crop. Azolla as dual crop in crop estimate to reduce nitrogen requirement by 20–25% [21].

13. Conclusion

In developing countries, the most important challenge is to produce sufficient food for the growing population from inelastic land area. These microbes siphon out appreciable amounts of nitrogen from the atmospheric reservoir, solubilise phosphorus and enrich the soil with this important but scarce nutrient. The crop bacterial soil ecosystem can, therefore, be energized in sustainable agriculture with considerable ecological stability and environmental quality.

Author details

Prabakaran Elavarasi*, Muthuraman Yuvaraj and Pandurangan Gayathri
Adhiparasakthi Agricultural College, Kalavai, Tamil Nadu, India

*Address all correspondence to: elavarasi.praba94@gmail.com

References

[1] El-Gawad AAM, Hendawey MH, Farag HIA. Interaction between biofertilization and canola genotypes in relation to some biochemical constituents under Siwa Oasis conditions. Research Journal of Agriculture and Biological Sciences. 2009;**5**(1):82-96

[2] Bahadur AJ, Singh AK, Upadhyay RM. Effect of organic amendments and bio-fertilizers on growth, yield and quality attributes of Chinese cabbage (*Brassica pekinensis Olsson*). Indian Journal of Agricultural Research. 2006;**76**(10):596-598

[3] Beg MA, Singh JK. Effects of biofertilizers and fertility levels on growth, yield and nutrient removal of green gram (*Vigna radiata*) under Kashmir conditions. Indian Journal of Agricultural Sciences. 2009;**79**(5):388-390

[4] Martınez-Viveros O, Jorquera M, Crowley D, Gajardo G, Mora M. Mechanisms and practical considerations involved in plant growth promotion by rhizobacteria. Journal of Soil Science and Plant Nutrition. 2010;**10**:293-319

[5] Kang SM, Radhakrishnan R, You YH, Joo GJ, Lee IJ. Phosphate solubilizing *Bacillus megaterium* mj1212 regulates endogenous plant carbohydrates and amino acids contents to promote mustard plant growth. Indian Journal of Microbiology. 2014;**54**:427-433

[6] Kuan KB, Othman R, Abdul Rahim K, Shamsuddin ZH. Plant growth-promoting rhizobacteria inoculation to enhance vegetative growth, nitrogen fixation and nitrogen remobilisation of maize under greenhouse conditions. PLoS One. 2016;**11**(3):e0152478. DOI: 10.1371/journal.pone.0152478

[7] Kang SM, Radhakrishnan R, You YH, Joo GJ, Lee IJ, Lee KE. Phosphate solubilizing *Bacillus megaterium* mj1212 regulates endogenous plant carbohydrates and amino acids contents to promote mustard plant growth. Indian Journal of Microbiology. 2014;**54**:427-433

[8] Hayat R, Ali S, Amara U, Khalid R, Ahmed I. Soil beneficial bacteria and their role in plant growth promotion: A review. Annales de Microbiologie. 2010;**60**:579-598

[9] Ding Y, Wang J, Liu Y, Chen S. Isolation and identification of nitrogen-fixing bacilli from plant rhizospheres in Beijing region. Journal of Applied Microbiology. 2005;**99**:1271-1281

[10] Nadeem SM, Shaharoona B, Arshad M, Crowley DE. Population density and functional diversity of plant growth promoting rhizobacteria associated with avocado trees in saline soils. Applied Soil Ecology. 2012;**62**:147-154

[11] Walker EL, Connolly EL. Time to pump iron: Iron-deficiency signalling mechanisms of higher plants. Current Opinion in Plant Biology. 2008;**11**:530-535

[12] Kucera B, Cohn MA, Leubner-Metzger G. Plant hormone interactions during seed dormancy release and germination. Seed Science Research. 2005;**15**:281-307

[13] Bentsink L, Koornneef M. Seed dormancy and germination. In: The Arabidopsis Book. The American Society of Plant Biologists. 2008;**6**:e0119. DOI: 10.1199/tab.0119

[14] Hashem A, Abd Allah EF, Abdulaziz A, Radhakrishnan R, Kumar A. Plant defense approach of Bacillus subtilis (BERA 71) against *Macrophomina phaseolina* (Tassi) Goid in mung bean. Journal of Plant Interactions. 2017;**12**(1):390-401

[15] Lata C, Prasad M. The role of DREBs in the regulation of abiotic stress responses in plants. Journal of Experimental Botany. 2011;**62**:4731-4748

[16] Tiwari DD, Katiyar NK, Nigam RC, Gupta BR. Yield, nutrient uptake and quality characteristics of barley (*Hordeum vulgare L.*) as affected by graded levels of nitrogen and Azotobacterinoculation. Research on Crops. 2008;**9**(2):243-245

[17] Marschner P, Crowley D, Rengel Z. Rhizosphere interactions between microorganisms and plants govern iron and phosphorus acquisition along the root axis–model and research methods. Soil Biology and Biochemistry. 2011;**43**:883-894

[18] Dursun A, Ekinci M, Donmez MF. Effects of foliar application of plant growth promoting bacterium on chemical contents, yield and growth of tomato (*Lycopersicon esculentum L.*) and cucumber (*Cucumis sativus L.*). Pakistan Journal of Botany. 2010;**42**:3349-3356

[19] Singh A, Tripathi PN, Singh R. Effect of rhizobium inoculation, nitrogen and phosphorus levels on growth, yield and quality of Kharif cowpea (*Vigna ungiculataH. Walp*). Crop Research. 2007;**33**(1-3):71-73

[20] Xie S, Wu HJ, Zang H, Wu L, Zhu Q, Gao X. Plant growth promotion by spermidine-producing *Bacillus subtilis* OKB105. Molecular Plant-Microbe Interactions. 2014;**27**:655-663

[21] Barnawal D, Maji D, Bharti N, Chanotiya CS, Kalra A. ACC deaminase-containing Bacillus subtilis reduces stress ethylene-induced damage and improves mycorrhizal colonization and rhizobial nodulation in Trigonella foenum-graecum under drought stress. Journal of Plant Growth Regulation. 2013;**32**:809-822

Chapter 3

Applications and Constraints of Plant Beneficial Microorganisms in Agriculture

Sovan Debnath, Deepa Rawat, Aritra Kumar Mukherjee, Samrat Adhikary and Ritesh Kundu

Abstract

At present time, chemical fertilizers are more in practice for crop production, which failed to upkeep soil and environment quality and affected the sustainability of the agricultural production system. Conversely, biofertilizers are ecosystem friendly, one of the best modern tools for agriculture, and are used to improve soil fertility and quality. Biofertilizers have now emerged as a highly potent alternative to inorganic fertilizers and offer an ecologically sound and economically attractive route for augmenting nutrient supply and increasing crop production. These include live cells of diverse genera of microorganisms and have the potential to fix atmospheric nitrogen and solubilize and mobilize plant nutrients from insoluble form through microbiological process. It has also the potential to diminish the gap between nutrient supply through fertilizers and nutrient removal by crops. Hence, biofertilizers can be a feasible option to the farmers to increase crop productivity and should find greater acceptance from the extension workers and commercial biofertilizer manufacturers.

Keywords: N fixers, P-K mobilizers, biofertilizer formulation, current advances

1. Introduction

Biofertilizers, more appropriately microbial inoculants, are the preparations containing one or more species of microorganisms which have the ability to capture or mobilize nutritionally important plant nutrients from non-usable to usable form through the biological processes such as N fixation, P solubilization, excretion of plant growth enhancers, or cellulose degradation in soil, compost, and other environments [1–3]. Biofertilizers are low-cost and environment-friendly supplement to chemical fertilizers and manures. Recently, biofertilizers are gaining momentum due to its ability to maintain soil health, minimize environmental degradation, and cut down the use of inorganic fertilizers in agriculture. These inputs gained added importance in rainfed agriculture in view of their low cost, as small to marginal farmers across the globe cannot afford expensive chemical fertilizers [4]. Biofertilizers could be an ideal input for cutting the cost of production and for practicing organic and conservation farming [5]. These organisms can be engaged in maintaining long-term soil fertility and sustainability [6, 7]. For the generations to come, biofertilizers are indispensable to ensure healthy soils and food.

The emphasis on chemical fertilizers, which sometimes led to unscientific and non-judicious application, has meant that the soil be regarded as an inert substrate for plant roots, instead of a living biosphere, the rhizosphere, containing a myriad of organisms [3]. The blanket use of inorganic fertilizers has also led to pollution of the soils and surface water bodies in many regions of the world [5]. Nevertheless, the importance of fertilizers, essential for achieving increased crop production, will further increase because there is little scope for bringing more areas under cultivation and majority of soils are deficient in many macro- and micronutrients. It is now realized that in agricultural lands under intensive monoculture system, including rice, which receives heavy application of chemical fertilizers alone, productivity slowly is declining, and environmental quality is deteriorating [8]. Intensification of agriculture has also widened the gap between nutrient removal and supplies and, thus, soil fertility depletion [9]. The role of biofertilizers in agriculture, therefore, assumes special significance, particularly in the present context of increased cost of inorganic fertilizers and their hazardous effects on soil health. The success with biofertilizers is reported for more than 100 years in many parts of the world, and statistically significant increase in yields has been observed [2]. However, their response varies with crops, host cultivars, locations, seasons, agronomic practices, bacterial strains, soil fertility, and interaction with native soil microflora.

2. Types of biofertilizers

Biofertilizers may be broadly classified into nitrogen-fixing bacteria, phosphate-solubilizing microorganisms, and organic matter decomposers (**Figure 1**). Nevertheless, it also includes organic fertilizers (manure, etc.), which are rendered in an available form due to the interaction of microorganisms or due to their association with plants.

2.1 Nitrogen-fixing biofertilizer (NFB)

Nitrogen-fixing organisms are used in biofertilizer as a living fertilizer composed of microbial inoculants or groups of microorganisms which are able to fix

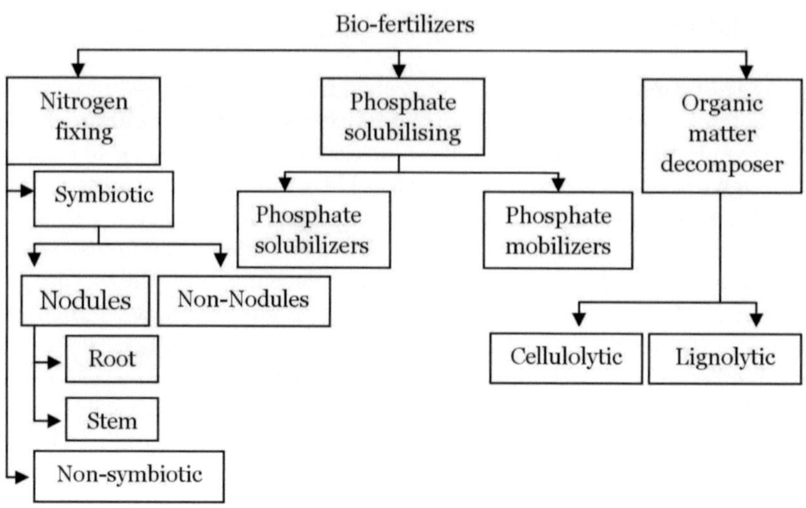

Figure 1.
A broad classification of biofertilizers.

atmospheric nitrogen, which is transformed into organic nitrogenous compound. The nitrogen-fixing bacteria work under two conditions, symbiotically (*Rhizobium*, *Frankia*) and as free-living bacteria (nonsymbiotic) such as *Azotobacter* and *Azospirillum*. The N_2-fixing bacteria associated with nonlegumes include species of *Achromobacter*, *Alcaligenes*, *Arthrobacter*, *Acetobacter*, *Azomonas*, *Beijerinckia*, *Bacillus*, *Clostridium*, *Enterobacter*, *Erwinia*, *Derxia*, *Desulfovibrio*, *Coryne bacterium*, *Campylobacter*, *Herbaspirillum*, *Klebsiella*, *Lignobacter*, *Mycobacterium*, *Rhodospirillum*, *Rhodopseudomonas*, *Xanthobacter*, *Mycobacterium*, and *Methylosinus* [10].

2.1.1 Symbiotic

The most exploited symbiotic N_2-fixing bacteria are those belonging to the family *Rhizobiaceae*. *Rhizobium* inoculants are of greatest importance because of their ability to fix atmospheric N_2 in association with certain legumes [11]. It is estimated that N_2 fixation by *Rhizobium* in root nodules of legumes is of the order of 14 million tons on a global scale and is almost 15% of the industrial N fixation. Yield of many legumes can be increased substantially by the use of appropriate *Rhizobium* cultures. For successful nodulation each legume requires a specific species of *Rhizobium* to form effective nodules. Many legumes may be modulated by diverse strains of rhizobia, but growth is enhanced only when nodules are produced by effective strains of rhizobia [12]. *Rhizobium* can be used for legumes crop and trees (e.g., lucerne) and is a crop-specific inoculant, for example, *Rhizobium trifolii* for berseem, *Rhizobium meliloti* for lucerne, *Rhizobium phaseoli* for green gram and black gram, *Rhizobium japonicum* for soya bean, *Rhizobium leguminosarum* for pea and lentil, *Rhizobium lupini* for chickpea, and *Rhizobium* spp. for cowpea. *Rhizobium* is however limited by cross-inoculation group, and only certain legumes are benefited by this symbiosis.

Similar to the *Rhizobium*, other filamentous bacteria of genus *Frankia* belonging to the family *Frankiaceae* are found in the root nodules of nonlegumes such as trees and shrubs. These bacteria live in symbiosis with actinorhizal plants. These actinorhizal plants are used for timber and fuel wood production, for wind breaks, and for shelterbelts in coastlines and desert, as well as for land reclamation [13]. In arid areas where actinorhizal plants are not present, inoculation of *Frankia* (*Frankia alni*) can be advantageous [13]. Despite their potential importance, very limited information is available for inoculation practice and their use for *Frankia* symbiosis. However, their potential could be harnessed in agroforestry system.

2.1.2 Nonsymbiotic

In nonsymbiotic or free-living nitrogen, fixation does not require host plant, and bacteria do not form nodules. An example of such free-living bacteria is *Azotobacter*. They fix atmospheric N_2 nonsymbiotically, and the extent of fixation is directly depends upon the amount of carbohydrates utilized by them [14, 15]. Azotobacter comprises seven species: *A. chroococcum*, *A. vinelandii*, *A. beijerinckii*, *A. paspali*, *A. armeniacus*, *A. nigricans*, and *A. salinestri* [16]. Soils containing poor organic matter and antagonistic relationship with other soil microorganism adversely affect the population of *Azotobacter*. Besides nitrogen fixation, it can also synthesize growth-promoting substances, viz., auxins, gibberellins, and to some extent the vitamins. It also helps to improve seed germination and crop growth due to positive response of B vitamins, naphthalene acetic acid (NAA), gibberellic acid (GA), and chemical produced during the biochemical process showing antagonistic relationship with root pathogen [17].

2.1.3 Associative

Apart from symbiotic and nonsymbiotic nitrogen fixers, some bacteria form a close associative symbiosis with the higher plants. These bacteria live on the root surface and sometimes also penetrate into the root tissues but do not produce any visible nodule or outgrowth on the root tissue. *Acetobacter diazotrophicus* and *Herbaspirillum* spp. associated with sorghum, maize, and sugarcane [18–20] and *Azospirillum*, *Bacillus*, *Enterobacter*, *Herbaspirillum*, *Klebsiella*, *Pseudomonas*, and *Rhizobium* associated with rice and maize [21] are examples of associative nitrogen-fixing microorganism.

Azospirillum produces growth-regulating substances, which help to protect from soilborne diseases. It improves leaf area index and ultimately crop yield. Apart from many species across the globe, the major species under this genus are *A. lipoferum* and *A. brasilense*. *Azospirillum* species mainly identified as rhizosphere bacteria and its colonization of the rhizosphere have been studied extensively [22–24]. *Azospirillum* with the plant having C_4-dicarboxylic pathway (Hatch and Slack pathway) of photosynthesis formed associative symbiosis because they fix nitrogen in salts of organic acids such as malic and aspartic acid [25]. So, it is mainly beneficial for C_4 plants like maize, sorghum, sugarcane, etc. Despite all these benefits that bear great promise as a growth-promoting N_2-fixing biofertilizer, the main problem that limits the use of *Azospirillum* is great uncertainty and unpredictability of the results [26].

2.1.4 Cyanobacteria

Blue green algae (BGA) are known as cyanobacteria. Cyano means blue, so that means it is blue bacteria. These belong to eight different families, phototrophic in nature, and produce auxins, indole acetic acid (IAA), and GA. N-fixing blue green algae have been shown to be the most important in maintaining and improving the productivity of rice fields [27]. Favorable condition for biological nitrogen fixation by BGA is considered to be one of the reasons for relatively stable yield of rice under flooded condition. BGA forms symbiotic association capable of fixing nitrogen with fungi, fern, and flowering plants, but the most common symbiotic association has been found between a free floating aquatic fern, the *Azolla* and the *Anabaena azollae* (BGA) [28]. This association produces 40–60 tons of organic matter per hectare per year. Despite the importance of N_2-fixing cyanobacteria in rice cultivation, the production and application are poorly developed. Biofertilizers should be seriously considered for supporting sustainable agriculture practice [29].

2.1.5 Azolla

Azolla is known as free floating water fern that fixes atmospheric N_2 in symbiotic association with BGA (*Anabaena azollae*) in rice field. They are free-living organism and use energy derived from photosynthesis to fix nitrogen. It is a fast-growing water fern and can double its weight within a week [30]. The most common species occurring in India is *A. pinnata*. *Azolla* is rich organic manure and mineralizes soil nitrogen rapidly which can be available to the crop in a very short period. *Azolla* can help rice or other crops through dual cropping or green manuring of soil [31].

2.2 Phosphate-solubilizing biofertilizer (PSB)

Several experiments have showed the ability of different bacterial species to solubilize insoluble inorganic phosphate minerals, such as tricalcium phosphate, dicalcium phosphate, hydroxyapatite, and rock phosphate. Phosphate-solubilizing bacteria are common in the rhizosphere, and secretion of organic acids like citric, oxalic, tartaric, acetic, lactic, gluconic, glyoxylic, maleic, and fumaric helps to convert insoluble form of phosphorus to plant available form [32]. Some of the bacterial genera are *Achromobacter*, *Agrobacterium*, *Micrococcus*, *Enterobacter*, and *Erwinia*. Among the soil bacterial communities, ectorhizospheric *Pseudomonas* and *Bacillus* and endosymbiotic rhizobia are found most effective phosphate solubilizers [33]. A higher amount of organic substances is present in the rhizosphere attracting the phosphate-solubilizing bacteria, and population is more in rhizospheric soil compared to the non-rhizospheric soil [34, 35]. Application of rock phosphate with PSB (*Bacillus megaterium* var. *phosphaticum*) showed that without phosphorus application PSB amendment could increase sugarcane yield up to 12.6% and it also improved sugar yield and juice quality [36]. Results of a greenhouse pot experiments with onion (*Allium cepa* L.) showed that application of *G. fasciculatum* along with *A. chroococcum* and 50% recommended P rate resulted in greater root length, plant height, bulb fresh weight, root colonization, and P uptake. Also the rate of chemical phosphatic fertilizer can be brought down [37]. Phosphate-solubilizing bacteria may be of greatest value in allowing the use of cheaper P sources.

2.3 Phosphate-mobilizing biofertilizer (PMB)

The symbiotic association between plant roots and fungi is termed as "mycorrhizal association." Arbuscular mycorrhizal fungi (AMF) form symbiotic relationship with about 90% of land plant species [38]. These are of two types, ectomycorrhiza found in trees and found beneficial for forest trees, and endomycorrhiza for crop plants [39]. The functional symbiosis in mycorrhizal fungus is obligatory and depends on host photosynthates and energy. The plant acquires carbon for various mycorrhizal benefits to the host plant. The fungi capture nutrients from soil solution with the help of mycelium that extends from the root surfaces into the soil matrix. So, it results more efficient nutrient uptake and improved plant growth when mycorrhizal fungi colonized the root systems [40].

In higher plants, phosphorus and other nutrients are often mediated with mycorrhizal association, in which symbiotic association is performed by higher plants and associative fungi (*Glomus*) [41]. Hyphae of AMF do not solubilize the insoluble unavailable phosphorus but assimilate them from soil for their own requirement. Mycorrhizal roots can take up several times more phosphorus per unit root length than non-mycorrhizal roots. Mycorrhizal symbiosis also increased the tolerance of heavy metal contamination or drought, as well as lesser susceptibility of root pathogens. AMF also helps to improve soil quality by having a direct influence on soil aggregation [42]. This association is generally found very effective in agroforestry. The other crops benefited from AMF are sorghum, barley, wheat, tobacco, cotton, soybean, apple, citrus, grape, etc.

2.4 Organic matter decomposer

Composting is a key technology to use different types of organic wastes (crop residues, rural and urban wastes), and it takes about 4–6 months for its maturity for use as a source of plan nutrients. To decompose these organic waste, some cellulolytic and lignolytic microorganisms are introduced which help to decompose that organic wastes at a faster rate and make it ready for use within 2–3 months. Many soilborne fungal species like *Aspergillus niger*, *Penicillium*, *Trichoderma viride*, *Trichurus spiralis*, *Phanerochaete chrysosporium*, etc. act as an activator in the decomposition process of plant bodies containing cellulose or lignin [43].

2.5 Potassium-solubilizing biofertilizer (KSB)

Some soil microorganisms are capable of solubilising potassium from K-bearing minerals such as muscovite, mica, orthoclase and illite. These minerals are the potential source of available K in soil. Microorganism produces organic substances which react with these K bearing minerals to solubilize K and enhances available K in the soil solution [44]. These organisms also produce various types of amino acids, growth-promoting compounds (IAA, GA, etc.), and vitamins, promoting the crop growth and yield [45]. *Frateuria aurantia*, a K-solubilizing bacteria, is capable of mobilizing mixture of potassium from mica into a usable form for the plants, which has fairly been applied to crops in association with other biofertilizers without any antagonistic effects [46, 47]. Application of high-K-bearing clay mineral with K-solubilizing bacteria can help to mitigate the K requirement in agricultural soils [48].

2.6 Sulfur-solubilizing biofertilizer (SSB)

Sulfur is one of the major elements in oil seed crops and some vegetables (onion, oat, cauliflower, etc.) and some species (ginger, garlic, etc.). It is essential for biochemical synthesis of some important glycosides, pungent compound, and disease resistance properties. Khandkar et al. [49] observed that the nodule in black gram was increased due to sulfur application. Deficiency of sulfur in agricultural soils could be corrected by application *Azotobacter pasturianam* as biofertilizer [50].

2.7 Zinc-solubilizing biofertilizer (ZSB)

Zinc is one of the micronutrients whose deficiency affects the crop growth and crop yield [5, 8]. Zinc fertilizers are very costly and its availability is also limited. So, zinc solubilizers can play a vital role for providing adequate supply of zinc to the crop and enhancing the crop growth and yield. The microorganisms which are well known for solubilization of zinc are *Bacillus subtilis*, *Thiobacillus thiooxidans*, and *Saccharomyces* sp. [51]. These strains are used as zinc biofertilizers and get positive response to the crop. Sometime application of zinc fertilizers combination with zinc biofertilizers (*Bacillus* sp.) gave better response and increased zinc concentration in the soil [46].

2.8 Plant growth-promoting rhizobacteria (PGPR)

Plant growth promoting rhizobacteria (PGPR), when grown in association with host plant, result in stimulation of growth of their host. It represents a wide variety of soil bacteria. These bacteria vary in their mechanism of plant growth promotion but generally influence growth via P solubilization, nutrient uptake enhancement, and plant growth hormone production [33, 52, 53]. Bertrand et al. [54] showed that a rhizobacterium belonging to the genus *Achromobacter* could enhance root hair number and length in rapeseed. The PGPR inoculants promote growth by any of the following mechanism: (i) suppression of plant disease (bioprotectants), (ii) improved nutrient acquisition (biofertilizers), and (iii) phytohormone production (biostimulants).

3. Potential of biofertilizers

The competent strains of nitrogen-fixing, phosphate-solubilizing, or cellulolytic microorganisms are used for application in seed, soil, and roots of saplings or composting areas with the intention to amplify the number of such microorganisms and speed up those microbial processes which supplement the availability of nutrients that can be easily assimilated by plants (**Table 1**).

3.1 *Rhizobium*

They can fix nitrogen 50–100 kg/ha with legumes only. The symbiotic relationship between leguminous crops and *Rhizobium* is very important for crop production system. It has been proven to be useful for pulse legumes like chickpea, red gram, pea, lentil, black gram, oil seed legumes like soybean and groundnut, and forage legumes like berseem and lucerne [77]. The suitable strain is capable to increase the crop yield up to 10–35% since N is fixed at 40–200 kg/ha which is able to meet up to 80–90% of N need of the crop [46].

3.2 *Azotobacter*

The presence of this organism has been reported from the rhizosphere of various crop plants such as rice (*Oryza sativa* L.), maize (*Zea mays* L.), sugarcane (*Saccharum officinarum* L.), bajra (*Pennisetum glaucum* L.), vegetables, and plantation crops [78]. It can fix N up to 25 kg/ha under optimal conditions and increase yield up to 40–50% [5]. It has been observed that *Azotobacter* improved the seed germination and crop growth owing to the affirmative response of B vitamins, NAA, GA, and other chemicals produced during the biochemical process that exhibited antagonistic relationship with root pathogens [17].

3.3 *Azospirillum*

Apart from their nitrogen-fixing ability of about 20–40 kg/ha, they are also known to produce various growth-regulating substances. The *Azospirillum* form associative symbiosis with plants having the C_4-dicarboxylic pathway of photosynthesis (Hatch and Slack pathway), as they grow and fix nitrogen on salts of organic acids such as malic and aspartic acid [25]. Thus, *Azospirillum* is mostly recommended for C_4 plants like maize, sugarcane, sorghum, pearl millet, etc. [5].

3.4 *Azolla*

Azolla can fix 100–150 kg N/ha/year in rice fields along with *Anabaena* [79]. It can also be incorporated as green manure by adding in the fields prior to rice planting. The most widespread species in India is *A. pinnata* and can be reproduced on commercial scale by vegetative means. India has recently introduced some species of *Azolla* (*A. caroliniana*, *A. microphylla*, *A. filiculoides*, and *A. mexicana*) for their large biomass production [80].

3.5 Blue green algae (BGA)

In India, rice is one of the main staple food crops grown by farmers by using of BGA and *Azolla* as a plant nutrient provider. Generally, BGA has been reported to be able to supply 50–100 kg/ha nitrogen through biological N fixation, and in addition, it is also known to supply plant growth-promoting substances to crop under puddled condition [81].

Keeping in view the importance of biofertilizer for sustainability in agriculture sector, the government of India has also ensured the quality and production of biofertilizers under Section 3 of essential commodities, Act 1955. The government

Biofertilizer	Recommended crop	Effect	Reference
Nitrogen-fixing biofertilizers			
Rhizobium	Bean	Increased straw and grain yield, harvest index, and agronomic fertilizer use efficiency	Yanni et al. [55]
		Increased nodule dry weight and seed yield	Koskey et al. [56]
	Cowpea, common bean, peas, fenugreek	Increased vegetative growth parameters, shoot minerals, and yield	Arafa et al. [57]
	Pea	Increased mean seed yield	Abera and Abeba [58]
	Faba bean	Improved enzymatic activity in inoculated soil	Beshir et al. [59]
Bradyrhizobium	Pigeon pea	Induced improvement in nodule dry weight, plant biomass, and shoot N uptake	Youseif et al. [60]
Azotobacter	Mulberry	Increased trends in silk filament length, cocoon weight, shell weight, and shell ratio	Moorthi et al. [61]
	Pearl millet	Improved plant height, dry matter accumulation, no. of effective tillers, grain per ear, and grain and stover yield	Yadav et al. [62]
	Cauliflower	Increased morphological character and yield	Subedi et al. [63]
	Wheat	Enhanced grain yield	Mahato and Kafle [64]
Azospirillum brasilense	Maize	Increased plant growth and improved biochemical traits	Zeffa et al. [65]
	Wheat	Enhanced plant growth and increased root depth, fresh weight of roots and shoots, and nutrient use efficiency	Sayed et al. [66]
Azospirillum lipoferum	Foxtail millet	Improved seed weight, panicle, dry weight of shoot and root, total N content of shoot, and root and grain yield	Rao and Charyulu [67]
Cyanobacteria	Rice	Improved yield	Bhoosan et al. [68]
Azolla	Rice	Increased grain and straw yield	Mishra et al. [69]
	Rice	Reduction in weed emergence	Biswas et al. [70]
Phosphate-solubilizing biofertilizers			
Pseudomonas spp.	Chickpea	High nodulation and stimulation of plant growth	Malik and Sandhu [71]
Bacillus spp.	Amaranth	Improved nutrient use efficiency	Pandey et al. [72]
Aspergillus niger	Wheat	Improved growth and P uptake	Xiao et al. [73]

Biofertilizer	Recommended crop	Effect	Reference
Bacillus thuringiensis	Rice	Increased shoot length	David et al. [74]
Phosphate-mobilizing biofertilizers			
VAM	Jatropha	Reduced salt stress	Kumar et al. [75]
	Maize	Enhanced concentration of P in plant	Sudova and Vosatka [76]

Table 1.
Effect of biofertilizers on crop improvement.

has issued a fertilizer (control) amendment order (FCO), 2006, with the gazette notification, S.O. 391 (E), dated on March 24, 2006, for biofertilizer production. After coming into enforcement of this order, four biofertilizers came under the FCO, i.e., *Rhizobium*, *Azotobacter*, *Azospirillum*, and phosphate-solubilizing bacteria [82]. Though the effect of biofertilizers on the crop production is slow, they possess vast potential for meeting plant nutrient requirements and sustaining soil quality while curtailing the use of chemical fertilizers. The development of biofertilizers has paced up in the last 20 years, and phosphate-solubilizing bacteria (PSB) have been reported to be used most widely among the farming community [83, 84].

4. Role of biofertilizers in alleviating abiotic stress in plants

4.1 Salinity

The condition of soil salinity generally inhibits the crop growth. High concentration of salts imparts pessimistic effects on plant metabolism and growth owing to the osmotic stress and accumulation of Na^+ and Cl^- ions [85]. Salt stress is responsible for obliteration of the microbial communities and carbon cycling in the soil [86]. Several researchers have recommended various chemical, physical, and biological methods for improving crop growth and performance under salt-affected soils [87–89]. Apart from this, various other advancements, counting traditional breeding and genetic engineering, have also been tried to improve the salinity tolerance in plants. However, such intercessions have little success rate, owing to the intricacy of salinity tolerance and slight genetic variability among germplasm accessions [90]. Among these methods, the biological means of improving crop growth has identified some promising outcomes so far.

Several researches of recent past have suggested the efficiency of cyanobacteria for remediation of salt-affected soil in laboratory studies and field trials [91–95]. There have been a variety of suggested mechanisms involved in reclaiming the salt-affected soils and promotion of plant growth by cyanobacteria. Li et al. [96] suggested the nitrogen fixation, extracellular polymeric substance production, the accumulation of compatible solutes, plant growth hormone production, active export of ions through K^+/Na^+ channels and Na^+/H^+ antiporters, and defense enzyme productions as possible mechanisms for salt-affected soil remediation using cyanobacteria. Khalilzadeh et al. [97] suggested that enhanced grain filling speed, photosynthesis, plant water accumulation, and flag leaf salt accumulation were some plausible mechanisms for cycocel and PGR-induced salt tolerance shown in wheat plants under pot experiment. After investigating the salt stress and inoculation effect on nodulation and growth of forage cowpea (*Vigna unguiculata* cv. Baladi), Omara and Tamer [98] reported the alleviation of detrimental effects of salt stress by applying dual inoculation with tolerant *Bradyrhizobium* SARSRh3 + *Bradyrhizobium* SARS-Rh5 due to improvement in nodulation, growth dynamics, increase in K uptake, and reduced Na uptake in forage cowpea plants.

The use of bacterial inoculation, specifically, plant growth-promoting rhizobacteria (PGPR), has proved to be effective in improving plant stress tolerance. Several reports claimed that PGPR successfully improved growth of a wide range of agricultural crops under environmental stress conditions [99–104]. The PGPR are also known to use several mechanisms to sustain the plant growth under salt stress. Rhizobacteria trigger the plant antioxidant defense mechanism by modifying the key enzymes activity, viz., superoxide dismutase (SOD), peroxidase (POD), and catalase (CAT) that forage the overproducing reactive oxygen species (ROS) and ultimately defend the plants from salt toxicity [100, 105]. PGPR-inoculated plants have also been reported to have changes in their root architecture owing to the increased indole-3-acetic acid (IAA) level that facilitates the plants to take up more nutrients under salinity stress condition in soil [106, 107]. In a field trial, Kamaraj and Padmavathi [108] reported that the seeds treated with triple inoculation of biofertilizer such as *Rhizobium*, phosphate-solubilizing bacteria, and VAM at 600 gm/ha gave higher crop growth and seed yield parameters under saline stress condition.

The use of microorganisms as biofertilizers has also been reported to alleviate the effect of salinity on vegetables. The inoculation of seeds of various vegetables, such as tomato, pepper, bean, and lettuce, with PGPR has resulted in augmented root and shoot growth, dry weight, fruit, and seed yield and improved the resistance of plants to salt stress [109]. Mahmood et al. [110] revealed that PGPR and Si synergistically improved the salinity tolerance in mung bean. The use of arbuscular mycorrhiza (AM) has also been recorded to improve salt stress in tomato, onion, and lettuce [111–113].

4.2 Drought

Drought stress influences a range of growth parameters and stress-responsive genes in plants under the situation of stress. Inadequate quantity of water generally reduces the cell size and membrane integrity; create reactive oxygen species; and lowers down the crop productivity by promoting leaf senescence [114]. The plant-associated microbes possess a variety of mechanisms to deal with harmful impact of drought on plants and soil. Apart from the water content, these microbes also supply nutrients and provide favorable environmental conditions for the sustainable growth of plants. These microbes are known to encourage plant growth and development by various potential mechanisms which include:

a. Synthesis of various phytohormones such as IAA, cytokinins, and abscisic acid

b. Production of bacterial exopolysaccharides

c. Production of 1-aminocyclopropane-1-carboxylate (ACC) deaminase

d. Promoting systemic tolerance

The PGPR have the ability to produce plant hormones like IAA that encourage plant growth under stress condition. IAA is the most vigorous auxin that regulates the vascular tissue differentiation, adventitious and lateral root differentiation, cell division, and shoot development under drought stress [115]. The exopolysaccharides synthesized by microbes also enable certain plants to tolerate drought. Three drought-tolerant bacterial strains, viz., *Proteus penneri* (Pp1), *Pseudomonas aeruginosa* (Pa2), and *Alcaligenes faecalis* (AF3), inoculated in maize crop resulted in increased relative water content, protein, and sugar [116]. Sandhya et al. [117] have also reported the improved plant resistance against drought stress by the use of exopolysaccharide-producing bacteria. Under the stress environment, ACC is an immediate precursor of ethylene. The ACC deaminase produced by bacteria hydrolyzes ACC into ammonia and alpha-ketobutyrate [118]. Vardharajula et al. [119] have reported the decrease in antioxidant activity and enhanced production of proline, free amino acid, and sugar in plants with microbial inoculants under drought stress. The mycorrhizal inoculation in consortium with specific bacteria has also been recorded to improve plant growth, nutrient uptake, and relative water content to decrease the effect of drought. Ortiz et al. [120] revealed that the association of *Pseudomonas putida* and *Bacillus thuringiensis* reduced the stomal conductance and electrolyte leakage owing to the accumulation of proline in shoot and root.

Tomato (*Lycopersicon esculentum* Mill) cv. Anakha treated with phosphate-solubilizing bacteria (*Bacillus polymyxa*) was reported to secrete excess proline to resist the drought condition [121]. Giri et al. [122] studied the physiological response of peas (*Pisum sativum* L.) when inoculated with ACC deaminase bacteria *Variovorax paradoxus* 5C-2 under moisture stress and watering conditions. It was reported that the bacterial effects were more apparent and consistent in moisture stress condition. The AM fungal inoculation reduced the concentration of malondialdehyde and soluble protein in plant leaf and enhanced the activities of SOD, POD, and CAT, which ultimately led to the improved osmotic adjustment and drought tolerance of mycorrhizae citrus-grafting seedlings [123]. Inoculation of *Glomus versiforme* in citrus plants has also been reported to improve the osmotic status of the plant in drought condition owing to the enhanced levels of nonstructural carbohydrates, K^+, Ca^{+2}, and Mg^{+2}, which helped the plants to resist the drought condition [124]. Ruiz-Sanchez et al. [125] revealed the increase in photosynthetic efficiency and the antioxidative response of rice plant in drought stress after inoculation of arbuscular mycorrhiza.

Phosphate-solubilizing microorganisms have positively increased the plant growth and phosphorus absorption in maize resulting in increasing the efficiency of plant tolerance to drought stress conditions [126]. Inoculation of *Pseudomonas* spp. to basal plants under water stress improved their antioxidant and photosynthetic pigment content. *Pseudomonas* spp. were also found to have affirmative influence on the seedling growth and seed germination under water stress [127]. Chavoshi et al. [128] reported that phosphorus- and potassium-solubilizing bacterial consortium was able to increase biomass and important physiological traits in red bean under limited irrigation conditions. Li et al. [129] investigated the response of synergistic application of superabsorbent polymer (SAP) and biofertilizers (*Paenibacillus beijingensis* BJ-18 and *Bacillus* sp. L-56) on plant growth, including wheat and cucumber in drought stress. Both the biofertilizers amended with SAP were recorded to promote germination rate of seeds, plant growth, and soil fertility (urease, sucrose, and dehydrogenase activities). Moreover, the quantitative real-time PCR analysis revealed that biofertilizer + SAP significantly regulated the expression levels of genes involved in ethylene biosynthesis, stress response, salicylic acid, and transcription activation in plants in the drought stress condition.

5. Application and doses of biofertilizers

Biofertilizers are usually applied along with carrier material in order to enhance their efficacy. Khosro and Yousef [130] elucidated that the use of these microorganisms along with carrier material makes it possible for the users to handle them easily, facilitate their long-term storage, and augment their effectiveness. The biofertilizers are usually used as seed treatment in which the inoculant is mixed with water to make form of slurry and then mixed with seeds (**Table 2**). In this case, the carrier material is generally used as fine powder to get the tight coating of inoculant on the seed surface. For this purpose the use of adhesive, such as gum arabic, methyl ethyl cellulose, sucrose solutions, and vegetable oils, is recommended.

5.1 Seed treatment

The seed treatment of biofertilizer is done by suspending 200 g of biofertilizer in 300–400 mL of water and mixed tenderly with 10 kg of seeds using an adhesive-like acacia gum, jiggery solution, etc. Thereafter, the seeds are spread on a clean sheet/cloth under the shade to dry. The shade dried seeds should be sown within 24 hours.

Name of organism	Mode of action	Host crops for which used	Method of application	Rate of inoculant	Remarks
Rhizobium	Symbiotic N_2 fixation	Legumes like pulses, soybean, groundnut	Seed treatment	200 g per 10 kg seed	Leaves residual N in soil for the next crop
Azotobacter	Nonsymbiotic N_2 fixation	Cereals, millets, cotton, vegetable	Seed treatment	200 g per 10 kg seed	Also controls certain diseases
Azospirillum	Associative N_2 fixation	Nonlegumes like maize, barley, oat, sorghum, millet, sugarcane, rice, etc.	Seed treatment	200 g per 10 kg seed	Produces growth-promoting substances, can be applied to legumes as co-inoculant
Phosphate solubilizers	Phosphorus solubilization	Soil application for all crops	Seed treatment	200 g per 10 kg seed	Can be mixed with rock phosphate
Blue green algae (BGA)	Nonsymbiotic N_2 fixation	Rice	Soil application	10 kg/ha	Reduces soil alkalinity, has growth-promoting effects
Azolla	Symbiotic N_2 fixation	Rice	Soil application	1 ton dried material/ha	—
Mycorrhiza (VAM)	Symbiotic association	Many tree species, wheat, sorghum, ornamentals	Soil application	—	Usually seedlings are inoculated

Table 2.
Application and doses of biofertilizers for various crops [43].

5.2 Seedling root dip

This method is generally applied for transplanted crops. For rice crop, a bed filled with water is prepared in the field, and recommended biofertilizers are mixed in this water. The roots of seedlings are dipped for 5–10 min and then transplanted.

5.3 Soil treatment

Four kilograms of the recommended biofertilizer is mixed in 200 kg of compost and kept overnight. This mixture is then incorporated in the soil at the time of sowing or planting.

5.4 Liquid biofertilizers

Bhattacharyya and Kumar [131] stated that biofertilizers manufactured in India are mostly carrier based and the microorganisms have the shelf life of only 6 months. The advantage of liquid biofertilizer over powder based is that microorganisms have longer shelf life up to 2 years and they are tolerant to UV radiations and high temperature (55°C). The count is as high as 109 c.f.u/ml, which is maintained constant up to 2 years. Since they are liquid formulation, the application in the field is very easy and simple. They are applied using hand sprayer, power sprayer, and fertigation tanks and as basal manure mixed along with farm yard manure (FYM) [132, 133].

For all leguminous crops, *Rhizobium* is generally applied as seed inoculant. *Azospirillum*/*Azotobacter* is inoculated through seed, seedling root dip, and soil application methods in transplanted crops. For direct sown crops, *Azospirillum* is usually incorporated through seed treatment or soil application.

6. Constraints in biofertilizer use

Despite little investment, eco-friendly character, and advantages of biofertilizers, adoption of this organic input by farmers has remained far from satisfactory. There are several constraints at production, marketing, and field level which limit the adoption of biofertilizers among the wide community of farmers.

6.1 Production constraints

- **Raw material:** Biofertilizers are generally prepared as carrier-based inoculants with effective microorganisms. Granular form of carrier material like peat, perlite, charcoal, etc. is commonly recommended for soil inoculation of the biofertilizer [46]. These carrier materials for seed and soil treatment are not easily available and accessible to the small and marginal farmers. In India, these carriers are neither available in adequate quantities nor in desirable quality, which is one of the reasons for the lack of popularity of biofertilizers among the Indian farmers [134].

- **Specificity of strains for different agroclimatic regions:** The majority of the strains of biofertilizers is not only crop specific but is also soil and agroclimate specific. The lack of region-specific strains is one of the major constraints associated with biofertilizer use. This confines their extensive and optimum use with expected performance [46, 135].

- **Biological constraints:** There is likelihood of presence of ineffective or antagonistic strains in the bio-inoculants, and removal of these strains from the bio-inoculant is generally a complicated task. The selected strains should also have the ability to compete with other strains, N-fixing or nutrient-solubilizing/nutrient-mobilizing ability over a range of environmental conditions, and ability to survive in broth and in inoculants carrier [134, 136]. This largely affects the efficiency of desired microorganism as biofertilizer.

- **Technical constraints:** Biofertilizers possess the tendency to mutate during fermentation which increases the cost of production and quality control. A broad range of research is needed to reduce such undesired changes [5].

- **Economic constraints:** For the production of quality product, the use of high-tech instruments and equipment is required. In the absence of these facilities, production of contamination free product is uncertain. Moreover, the lack of trained human resources in the production units and lack of suitable training on the production techniques also serve as a limitation of the widespread use of biofertilizers [137].

6.2 Marketing constraints

- **Limited transportation and storage facilities:** The serviceable life of biofertilizers prepared with common carriers like peat or lignite is usually less than 6 months. It has been recommended that best results of biofertilizers are possible only if the material is used within 3–4 months of production. But often the biofertilizers are subjected to very high temperature during transportation and storage which reduces their efficiency and leads to lack of interest among the dealers due to nominal profit margin [138, 139].

- **Low demand:** Owing to the lack of adequate promotion and awareness about the advantages of biofertilizers, farmers refrain themselves from adopting this sustainable practice due to different methods of inoculation and no visual variation in the crop growth immediately as in the case of inorganic fertilizers [46].

6.3 Field-level constraints

- Soil conditions like acidity, presence of salts and toxic elements, application of pesticides, water logging and drought [140]

- Poor organic matter content of many soils around the world

- Extreme annual and diurnal variation in soil temperature

- Poor competition and adaptability as compared to native soil microflora [46]

7. Conclusion

Enhancing agricultural crop production needs to be ushered through new horizons without causing any harm to the natural resources and environmental quality. So, low-cost and eco-friendly biofertilizers could play a critical role in increasing crop yield by cutting the use of chemical fertilizers and increased nutrient use efficiency vis-à-vis maintaining long-term soil fertility and quality. However, lack of consistent responses in different soils and environmental conditions, difficulties in application, limited shelf life, and slow action are reasons restraining the widespread commercialization of biofertilizers. We need to apprehend that biofertilizers are extremely specific to crops, soils, and edaphic factors and their sustainability in soils largely depends on pH, soil organic matter, native microbiota, and soil moisture and temperature regime. Our understanding on particular strain effectiveness with specific to crop, soil, and climate needs to be strengthened through extensive research and development. Research should also focus on standardizing biofertilizer dose in a particular soil and crop. Efforts from the government should be emphasized on frequent monitoring of the biofertilizer manufacturing units to assure proper method of production and top quality of the produce and storage. Wide publicity and large-scale utilization of this new era technology through research institutions, nongovernment organizations (NGOs), scientific training, farmer fairs or exhibitions, extension workers, and media are urged.

Author details

Sovan Debnath[1*], Deepa Rawat[2], Aritra Kumar Mukherjee[3], Samrat Adhikary[3] and Ritesh Kundu[3]

1 ICAR-Central Institute of Temperate Horticulture, Regional Station, Mukteshwar, Uttarakhand, India

2 College of Forestry, VCSG Uttarakhand University of Horticulture and Forestry, Ranichauri, Uttarakhand, India

3 Department of Agricultural Chemistry and Soil Science, Faculty of Agriculture, Bidhan Chandra Krishi Viswavidyalaya, Mohanpur, West Bengal, India

*Address all correspondence to: sovan.dta@gmail.com

References

[1] Vessey VK. Plant growth promoting rhizobacteria as biofertilizers. Plant and Soil. 2003;**255**:571-586

[2] Trimurtulu N, Rao DLN. Liquid Microbial Inoculants and their Efficacy on Field Crops. Amaravathi: ANGRAU, Agricultural Research Station; 2014. p. 54

[3] Pal S, Singh HB, Farooqui A, Rakshit A. Fungal biofertilizers in Indian agriculture: Perception, demand and promotion. Journal of Eco-friendly Agriculture. 2015;**10**:101-113

[4] Barman M, Paul S, Choudhury AG, Roy P, Sen J. Biofertilizers as prospective input for sustainable agriculture in India. International Journal of Current Microbiology and Applied Sciences. 2017;**6**:1177-1186

[5] Singh M, Dotaniya ML, Mishra A, Dotaniya CK, Regar KL, Lata M. Role of biofertilizers in conservation agriculture. In: Bisht JK et al., editors. Conservation Agriculture. Singapore: Springer; 2016. pp. 113-134

[6] Sharma A, Kumar S, Ahmed N, Sharma OC, Debnath S. Protection of fruit growing soils from the threats on the mark. In: Sharma A et al., editors. Soil Management Strategies-Thriving to Optimise Fruits Production. New Delhi: Kalyani Publishers; 2015. pp. 158-170

[7] Shelat HN, Vyas RV, Jhala YK. Biofertilizers and PGPR for evergreen agriculture. In: Verma DK, Srivastav PP, editors. Microorganisms in Sustainable Agriculture, Food and the Environment. Boca Raton: CRC Press; 2017

[8] Debnath S, Pachauri SP, Srivastava PC. Improving use efficiency of applied phosphorus fertilizer by zinc fertilization in basmati rice-wheat cropping system. Indian Journal of Agricultural Research. 2015;**49**:414-420

[9] Gupta N, Debnath S, Sharma S, Sharma P, Purohit J. Role of nutrients in controlling the plant diseases in sustainable agriculture. In: Meena VS et al., editors. Agriculturally Important Microbes for Sustainable Agriculture. Singapore: Springer Nature; 2017. pp. 217-262

[10] Wani SP. Inoculation with associative nitrogen fixing bacteria: Role in cereal grain production improvement. Indian Journal of Microbiology. 1990;**30**:363-393

[11] Sheraz MS, Hassan GI, Samoon SA, Rather H, Showkat A, Dar A, et al. Biofertilizers in organic agriculture. Journal of Phytology. 2010;**2**:42-54

[12] Cooper J. Multiple responses of rhizobia to flavonoids during legume root infection. In: Callow JA, editor. Advances in Botanical Research: Incorporating Advances in Plant Pathology. Cambridge: Academic Press; 2004. pp. 1-62

[13] Schwencke J, Caru M. Advances in actinorhizal symbiosis: Host plant-*Frankia* interaction, biology, and applications in arid land reclamation: A review. Arid Land Research and Management. 2001;**15**:285-327

[14] Stephens JHG, Rask HM. Inoculant production and formulation. Field Crops Research. 2000;**65**:249-258

[15] Gomare KS, Mese M, Shetkar Y. Isolation of *Azotobacter* and cost effective production of biofertilizer. Indian Journal of Applied Research. 2013;**3**:54-56

[16] Becking J. The family Azotobacteraceae. PRO. 2006;**6**:759-783

[17] Mazid M, Zeba HK, Quddusi S, Khan TA, Mohammad F. Significance of Sulphur nutrition against metal induced

oxidative stress in plants. Journal of Stress Physiology & Biochemistry. 2011;**7**:165-184

[18] Triplett E. Diazotrophic endophytes: Progress and prospects for nitrogen fixation in monocots. Plant and Soil. 1996;**186**:29-38

[19] James EK, Olivares FL, Baldani JI, Dobereiner J. *Herbaspirillum*, an endophytic diazotroph colonizing vascular tissue in leaves of *Sorghum bicolour* L. Moench. Journal of Experimental Botany. 1997;**48**:785-797

[20] Boddey RM, Da Silva LG, Reis V, Alves BJR, Urquiaga S. Assesment of bacterial nitrogen fixation in grass species. In: Triplett EW, editor. Prokaryotic Nitrogen Fixation: A Model System for Analysis of a Biological Process. Wymondham, UK: Horizon Scientific Press; 2000. pp. 705-726

[21] James EK. Nitrogen fixation in endophytic and associative symbiosis. Field Crops Research. 2000;**65**:197-209

[22] Burdman S, Kigel J, Okon Y. Effect of *Azospirillum brasilense* on nodulation and growth of common bean (*Phaseolus vulgaris* L.). Soil Biology and Biochemistry. 1997;**29**:923-929

[23] Pereg-Gerk L, Gilchrist K, Kennedy IR. Mutants with enhanced nitrogenase activity in hydroponic *Azospirillum brasilense* wheat association. Applied and Environmental Microbiology. 2000;**66**:2175-2184

[24] Steenhoudt O, Vanderlayden J. *Azospirillum*, a free living-nitrogen fixing bacteria closely associated with grasses: Genetic, biochemical and ecological aspects. FEMS Microbiology Reviews. 2000;**24**:487-506

[25] Arun KS. Bio-Fertilizers for Sustainable Agriculture. Mechanism of P-Solubilization. Sixth ed. Jodhpur, India: Agribios Publishers; 2007. pp. 196-197

[26] Van de Broek A, Dobbelaere S, Vanderleyden J, Vandommelen A. *Azospirillum*-plant root interactions: Signaling and metabolic interactions. In: Triplett EW, editor. *Prokaryotic Nitrogen Fixation: A Model System for Analysis of a Biological Process*. Wymondham, UK: Horizon Scientific Press; 2000. pp. 761-777

[27] Watanabe I, Roger PA. Nitrogen fixation in wetland rice fields. In: Subba Rao NS, editor. Current Developments in Biological Nitrogen Fixation. Oxford: IBH, New Delhi; 1984. pp. 237-276

[28] Rahman MM, Amano T, Shiraiwa T. Nitrogen use efficiency and recovery from N fertilizer under rice-based cropping systems. Australian Journal of Crop Science. 2009;**3**:336-351

[29] Hashem MA. Problem and prospects of cyanobacterial biofertilizers for rice cultivation. Australian Journal of Plant Physiology. 2001;**28**:881-888

[30] Kannaiyan S. Blue green algae biofertilizers. In: Kannaiyan S, editor. The Biotechnology of Biofertilizers for Rice Crops. Coimbatore, India: Tamil Nadu Agric. University publications; 1990. p. 225

[31] Ghosh N. Promoting biofertilisers in Indian agriculture. Economic and Political Weekly. 2004;**5**:5617-5625

[32] Kim KY, Jordan D, McDonald GA. Effect of phosphate solubilizing bacteria and vesicular–arbuscular mycorrhizae on tomato growth and soil microbial activity. Biology and Fertility of Soils. 1998;**26**:79-87

[33] Richardson AE. Prospects for using soil microorganisms to improve the acquisition of phosphorus by plants. Australian Journal of Plant Physiology. 2001;**28**:897-906

[34] Ahmad M, Nadeem SM, Naveed M, Zahir ZA. Potassium-solubilizing bacteria and their application in agriculture. In: Meena VS, Maurya BR, Verma JP, Meena RS, editors. Potassium Solubilizing Microorganisms for Sustainable Agriculture. New Delhi: Springer; 2016. pp. 293-313

[35] Bahadur I, Maurya BR, Kumar A, Meena VS, Raghuwanshi R. Towards the soil sustainability and potassium-solubilizing microorganisms. In: Meena VS, Maurya BR, Verma JP, Meena RS, editors. Potassium Solubilizing Microorganisms for Sustainable Agriculture. New Delhi: Springer; 2016. pp. 225-266

[36] Sundara B, Natarajan V, Hari K. Influence of phosphorus solubilizing bacteria on the changes in soil available phosphorus and sugar cane and sugar yields. Field Crops Research. 2002;**77**:43-49

[37] Carter OG. The effect of chemical fertilizers on seedling establishment. Animal Production Science. 1967;**7**:174-180

[38] Gadkar V, David-Schwartz R, Kunik T, Kapulnik Y. Arbuscular mycorrhizal fungal colonization factors involved in host reorganization. Plant Physiology. 2001;**127**:1493-1499

[39] Karandashov V, Bucher M. Symbiotic phosphate transport in arbuscular mycorrhizas. Trends in Plant Science. 2005;**10**:22-29

[40] Adholeya A, Tiwari P, Singh R. Largescale inoculums production of arbuscular mycorrhizal fungi on root organs and inoculation strategies. In: Declerck S, Strullu DG, Fortin A, editors. Soil Biology, Volume 4, *In vitro* Culture of Mycorrhizae. Berlin Heidelberg: Springer-Verlag; 2005. pp. 315-338

[41] Rillig MC, Wright SF, Eviner VT. The role of arbuscular mycorrhizal fungi and glomalin in soil aggregation: Comparing effects of the five plant species. Plant and Soil. 2002;**238**:325-333

[42] Kumar A, Patel JS, Bahadur I, Meena VS. The molecular mechanisms of KSMs for enhancement of crop production under organic farming. In: Meena VS, Maurya BR, Verma JP, Meena RS, editors. Potassium Solubilizing Microorganisms for Sustainable Agriculture. New Delhi: Springer; 2016. pp. 61-75

[43] Narayanasamy G, Arora BR, Biswas DR, Khanna SS. Fertilizers, manures and biofertilizers. In: Goswami NN et al., editors. Fundamentals of Soil Science. New Delhi: Indian Society of Soil Science; 2012. pp. 579-621

[44] Dotaniya ML, Meena VD, Basak BB, Meena RS. Potassium uptake by crops as well as microorganisms. In: Meena VS, Maurya BR, Verma JP, Meena RS, editors. Potassium Solubilizing Microorganisms for Sustainable Agriculture. New Delhi: Springer; 2016. pp. 267-280

[45] Sharma A, Shankhdhar D, Shankhdhar SC. Potassium-solubilizing microorganisms: Mechanism and their role in potassium solubilization and uptake. In: Meena VS, Maurya BR, Verma JP, Meena RS, editors. Potassium Solubilizing Microorganisms for Sustainable Agriculture. New Delhi: Springer; 2016. pp. 203-219

[46] Mazid M, Khan TA. Future of bio-fertilizers in Indian agriculture: An overview. International Journal of Agricultural and Food Research. 2014;**3**:10-23

[47] Raghavendra MP, Nayaka SC, Nuthan BR. Role of rhizosphere microflora in potassium solubilization. In: Meena VS, Maurya BR, Verma JP, Meena RS, editors. Potassium Solubilizing Microorganisms for

Sustainable Agriculture. New Delhi: Springer; 2016. pp. 43-59

[48] Samoon HA, Dar SA, Zehra B, Mahdi SS, Hassan SA. Bio-fertilizers in organic agriculture. Journal of Phytology. 2010;**2**:9-14

[49] Khandkar MR, Shinde DA, Kanekar VS, Jambley NR, Jain NK. Growth, nodulation and yield of rainfed black gram as influenced by phosphate and sulphate fertilization in vertisol. Indian Journal of Plant Physiology. 1985;**28**:318-322

[50] Saraf CS. Sulphur fertilization to pulses in yield and quality. In: Proc. TSIFAI Symp. Sulphur in Indian Agriculture. New Delhi; 1988. p. S II/2-1

[51] Raj SA. Bio-fertilizers for micronutrients. Biofertilizer Newsletters. 2007;**15**:8-10

[52] Bashan Y, Harrison SK, Whitmoyer RE. Enhanced growth of wheat and soybean plant inoculated with *Azopirillum brasilense* is not necessary due to general enhancement of mineral uptake. Applied and Environmental Microbiology. 1990;**56**:769-775

[53] Goldstein AH, Braverman K, Osorio N. Evidence for mutualism between a plant growing in a phosphate-limited desert environment and a mineral phosphate solubilizing (MPS) rhizobacterium. FEMS Microbiology Ecology. 1999;**30**:295-300

[54] Bertrand H, Plassard C, Pinochet X, Touraine B, Normand P, Cleyet Marel JC. Stimulation of the ionic transport system in *Brassica napus* by a plant growth-promoting rhizobacterium (*Achromobacter sp*.). Canadian Journal of Microbiology. 2000;**46**:229-236

[55] Yanni Y, Zidanb M, Dazzoc F, Rizka R, Mehesena A, Abdelfattaha F, et al. Enhanced symbiotic performance and productivity of drought stressed common bean after inoculation with tolerant native rhizobia in extensive fields. Agriculture, Ecosystems and Environment. 2016;**232**:119-128

[56] Koskey G, Mburu SW, Njeru EM, Kimiti MJ, Ombori O, Maingi. Potential of native rhizobia in enhancing nitrogen fixation and yields of climbing beans (*Phaseolus vulgaris* L.) in contrasting environments of eastern Kenya. Frontiers in Plant Science. 2017;**8**:443. DOI: 10.3389/fpls.2017.00443

[57] Arafa MM, El-Batanony NH, Nofal AM. Inoculation effect of rhizobial strains on growth, yield and chemical composition of some legume crops in new reclaimed soil. Middle East Journal of Agriculture Research. 2018;**7**:352-363

[58] Abera T, Abebe Z. Effects of fertilizer, rhizobium inoculation and lime rate on growth and yields field pea in horro and gedo highlands. Advances in Crop Science and Technology. 2018;**6**:397. DOI: 10.4172/2329-8863.1000397

[59] Beshir HM, Walley FL, Bueckert R, Taran B. Response of snap bean cultivars to rhizobium inoculation under dryland agriculture in Ethiopia. Agronomy. 2015;**5**:291-308

[60] Youseif SH, Abd El-Megeed FH, Saleh A. Improvement of faba bean yield using *Rhizobium/Agrobacterium* inoculant in low-fertility sandy soil. Agronomy. 2017;**7**:2. DOI: 10.3390/agronomy7010002

[61] Moorthi M, Senthilkumar A, Thangaraj A. The effect of biofertilizer *Azotobacter chrococcum* on the growth of mulberry crop *Morus indica* L. and the yield of *Bombyx mori* L. International Journal of Environment, Agriculture and Biotechnology. 2016;**1**:4-7

[62] Yadav PK, Verma R, Bamboriya JK, Yadav S, Jeeterwal RC. Response of

pearl millet *Pennisetum glaucum* L. to integrated nitrogen management. International Journal of Current Microbiology and Applied Sciences. 2019;**8**:429-437

[63] Subedi R, Khanal A, Aryal K, Chhetri LB, Kandel BP. Response of *Azotobacter* in cauliflower (*Brassica oleracea* L. *Var. Botrytis*) production at Lamjung, Nepal. Acta Scientifica Malaysia. 2019;**3**:17-20

[64] Mahato S, Kafle A. Comparative study of *Azotobacter* with or without other fertilizers on growth and yield of wheat in Western hills of Nepal. Annals of Agrarian Science. 2018;**16**:250-256

[65] Zeffa DM, Perini LJ, Silva MB, de Sousa NV, Scapim CA, Oliveira ALM, et al. *Azospirillum brasilense* promotes increases in growth and nitrogen use efficiency of maize genotypes. PLoS One. 2019;**14**:e0215332

[66] Sayed E, Hameda EA, Althubiani AS. Enhancement of plant growth by soil inoculation with *Azospirillum brasilense* HM1 isolated from soil of Saudi Arabia. International Journal of Current Microbiology and Applied Sciences. 2015;**4**:238-248

[67] Rao KVB, Charyulu PBBN. Evaluation of effect of inoculation of *Azospirillum* on the yield of *Setaria italica* (L.). African J. Biotech. 2005;**4**:989-995

[68] Bhooshan N, Pabbi S, Singh A, Sharma A, Chetan, Jaiswal A, et al. Impact of blue green algae (BGA) technology: An empirical evidence from northwestern indo-Gangetic Plains. Biotech. 2018;**8**:324. DOI: 10.1007/s13205-018-1345-5

[69] Mishra N, Patra AK, Garnayak LM, Mohanty AK, Swain SK. Long-term effects of integrated nutrient management on productivity and soil properties of rice (*Oryza sativa*)–rice cropping system in coastal Odisha. Indian Journal of Agronomy. 2017;**62**:239-246

[70] Biswas M, Parveen S, Shimozawa H, Nakagoshiw N. Effects of Azolla species on weed emergence in a rice paddy ecosystem. Weed Biology and Management. 2005;**5**:176-183

[71] Malik DK, Sindhu SS. Production of indole acetic acid by *Pseudomonas* sp.: Effect of coinoculation with *Mesorhizobium* sp. *Cicer* on nodulation and plant growth of chickpea (*Cicer arietinum*). Physiology and Molecular Biology of Plants. 2011;**17**:25-32

[72] Pandey C, Negi YK, Maheshwari DK, Rawat D, Prabha D. Potential of native cold tolerant plant growth promoting bacilli to enhance nutrient use efficiency and yield of *Amarathus hypochondriacus*. Plant and Soil. 2018;**428**:307-320. DOI: 10.1007/s11104-018-3681-y

[73] Xiao C, Zhang H, Fang Y, Chi R. Evaluation for rock phosphate solubilization in fermentation and soil–plant system using a stress-tolerant phosphate-solubilizing *Aspergillus niger* WHAK1. Applied Microbiology and Biotechnology. 2013;**169**:123-133

[74] David P, Raj RS, Linda R, Rhema SB. Molecular characterization of phosphate solubilizing bacteria (PSB) and plant growth promoting rhizobacteria (PGPR) from pristine soils. International Journal of Innovative Science, Engineering and Technology. 2014;**1**:317-324

[75] Kumar A, Sharma S, Mishra S. Influence of arbuscular mycorrhizal (AM) fungi and salinity on seedling growth, solute accumulation, and mycorrhizal dependency of *Jatropha curcas* L. Journal of Plant Growth Regulation. 2010;**29**:297-306

[76] Sudova R, Vosatka M. Differences in the effects of three arbuscular

mycorrhizal fungal strains on P and Pb accumulation by maize plants. Plant and Soil. 2007;**296**:77-83

[77] Saikia SP, Jain V. Biological nitrogen fixation with non-legumes: An achievable target or a dogma. Current Science. 2007;**92**:317-322

[78] Mazid M, Khan TA, Zeba HK, Quddusi S, Mohammad F. Occurrence, biosynthesis and potentialities of ascorbic acid in plants. International Journal of Plant, Animal and Environmental Sciences. 2011;**1**:167-184

[79] Mahajan A, Bhagat RM, Trikha A. Fertilizing bio-fertilizers. Agriculture Today. 2003a;**6**:36-37

[80] Yadav RK, Abraham G, Singh YV, Singh PK. Advancements in the utilization of Azolla-anabaena system in relation to sustainable agricultural practices. Proceedings of the Indian National Science Academy. 2014;**80**:301-316

[81] Mahajan A, Choudhary AK, Jaggi RC, Dogra RK. Importance of bio-fertilizers in sustainable agriculture. Farm Forum. 2003b;**3**:17-19

[82] Yadav AK. Glimpses of biofertilizers covered in fertilizer (control) order, 1985 (amendment, march 2006). Biofertilizer Newsletters. 2006;**14**:3-7

[83] Rao DLN. Microbial diversity, soil health and sustainability. Journal of the Indian Society of Soil Science. 2007;**55**:392-403

[84] Rao DLN. Recent advances in biological nitrogen fixation in agricultural systems. Proceedings of the Indian National Science Academy. 2014;**80**:359-378

[85] Deinlein U, Stephan AB, Horie T, Luo W, Xu G, Schroeder JI. Plant salt tolerance mechanisms. Trends in Plant Science. 2014;**19**:371-379

[86] Rath KM, Rousk J. Salt effects on the soil microbial decomposer community and their role in organic carbon cycling: A review. Soil Biology and Biochemistry. 2015;**81**:108-123

[87] Luo S, Tian L, Chang C, Wang S, Zhang J, Zhou X, et al. Grass and maize vegetation systems restore saline-sodic soils in the Songnen plain of Northeast China. Land Degradation and Development. 2018;**29**:1107-1119

[88] Saifullah, Dahlawi S, Naeem A, Rengel Z, Naidu R. Biochar application for the remediation of salt-affected soils: Challenges and opportunities. Science of the Total Environment. 2018;**625**:320-335

[89] Zhao Y, Wang S, Li Y, Liu J, Zhuo Y, Chen H, et al. Extensive reclamation of saline-sodic soils with flue gas desulfurization gypsum on the Songnen plain, Northeast China. Geoderma. 2018;**321**:52-60

[90] Krishna G, Singh BK, Kim EK, Morya VK, Ramteke PW. Progress in genetic engineering of peanut (*Arachis hypogaea* L.)— A review. Plant Biotechnology Journal. 2015;**13**:147-162

[91] Jaiswal P, Kashyap AK, Prasanna R, Singh PK. Evaluating the potential of *N. calcicola* and its bicarbonate resistant mutant as bioameleorating agents for 'usar' soil. Indian Journal of Microbiology. 2010;**50**:12-18

[92] Rady MM, Taha SS, Kusvuran S. Integrative application of cyanobacteria and antioxidants improves common bean performance under saline conditions. Scientia Horticulturae. 2018;**233**:61-69

[93] Roman JR, Roncero-Ramos B, Chamizo S, Rodriguez-Caballero E, Canton Y. Restoring soil functions by means of cyanobacteria inoculation: Importance of soil conditions and species selection. Land Degradation and Development. 2018;**29**:3184-3193

[94] Singh JS, Pandey VC, Singh DP. Efficient soil microorganisms: A new dimension for sustainable agriculture and environmental development. Agriculture, Ecosystems and Environment. 2011;**140**:339-353

[95] Singh V, Singh DV. Cyanobacteria modulated changes and its impact on bioremediation of saline-alkaline soils. Bangladesh Journal of Botany. 2015;**44**:653-658

[96] Li H, Zhao Q, Huanga H. Current states and challenges of salt-affected soil remediation by cyanobacteria. Science of the Total Environment. 2019;**669**:258-272

[97] Khalilzadeh R, Sharifia RS, Jalilianb J. Growth, physiological status, and yield of salt-stressed wheat (*Triticum aestivum* L.) plants affected by biofertilizer and cycocel applications. Arid Land Research and Management. 2017;**32**:71-90. DOI: 10.1080/15324982.2017.1378282

[98] Omara AD, Tamer G. Alleviation of salinity stress effects in forage cowpea (*Vigna unguiculata* L.) by *Bradyrhizobium* sp. inoculation. Microbiology Research Journal International. 2017;**23**:1-16

[99] Ji SH, Gururani MA, Chun SC. Isolation and characterization of plant growth promoting endophytic diazotrophic bacteria from Korean rice cultivars. Microbiological Research. 2014;**169**:83-98

[100] Islam F, Yasmeen T, Arif MS, Ali S, Ali B, Hameed S, et al. Plant growth promoting bacteria confer salt tolerance in *Vigna radiata* by up-regulating antioxidant defense and biological soil fertility. Plant Growth Regulation. 2015;**80**:23-36

[101] Majeed A, Abbasi MK, Hameed S, Imran A, Rahim N. Isolation and characterization of plant growth-promoting rhizobacteria from wheat rhizosphere and their effect on plant growth promotion. Frontiers in Microbiology. 2015;**6**:198. DOI: 10.3389/fmicb.2015.00198

[102] Rolli E, Marasco R, Vigani G, Ettoumi B, Mapelli F, Deangelis ML, et al. Improved plant resistance to drought is promoted by the root–associated microbiomes water stress–dependent trait. Environmental Microbiology. 2015;**17**:316-331

[103] Timmusk S, Kim SB, Nevo E, Abd ElDaim I, Ek B, Bergquist J, et al. Sfp-type PPTase inactivation promotes bacterial biofilm formation and ability to enhance wheat drought tolerance. Frontiers in Microbiology. 2015;**6**:387. DOI: 10.3389/fmicb.2015.00387

[104] Zahid M, Abbasi MK, Hameed S, Rahim N. Isolation and identification of indigenous plant growth promoting rhizobacteria from Himalayan region of Kashmir and their effect on improving growth and nutrient contents of maize (*Zea mays* L.). Frontiers in Microbiology. 2015;**6**:207. DOI: 10.3389/fmicb.2015.00207

[105] Jha Y, Subramanian R. PGPR regulate caspase-like activity, programmed cell death, and antioxidant enzyme activity in paddy under salinity. Physiology and Molecular Biology of Plants. 2014;**20**:201-207

[106] Vacheron J, Desbrosses G, Bouffaud ML, Touraine B, Moënne-Loccoz Y, Muller D, et al. Plant growth-promoting rhizobacteria and root system functioning. Frontiers in Plant Science. 2013;**4**:356. DOI: 10.3389/fpls.2013.00356

[107] Goswami D, Dhandhukia P, Patel P, Thakker JN. Screening of PGPR from saline desert of Kutch: Growth promotion in Arachis hypogeal by *Bacillus licheniformis* A2. Microbiological Research. 2014;**169**:66-75

[108] Kamaraj A, Padmavathi S. Alleviation of saline salt stress through pre-sowing biofertilizer seed treatment on crop growth and seed yield in green gram CV ADT3. Journal of Pharmacognosy and Phytochemistry. 2018;**7**(1):2205-2209

[109] Egamberdieva D, Lugtenberg B. Use of plant growth-promoting rhizobacteria to alleviate salinity stress in plants. In: Miransari M, editor. Use of Microbes for the Alleviation of Soil Stresses. New York, NY, USA: Springer; 2014. pp. 73-96

[110] Mahmood S, Daur I, Al-Solaimani SG, Ahmad S, Madkour MH, Yasir M, et al. Plant growth promoting rhizobacteria and silicon synergistically enhance salinity tolerance of mung bean. Frontiers in Plant Science. 2016;**7**:876

[111] Latef AAHA, Chaoxing H. Effect of arbuscular mycorrhizal fungi on growth, mineral nutrition, antioxidant enzymes activity and fruit yield of tomato grown under salinity stress. Scientia Horticulturae. 2011;**127**:228-233

[112] Cantrell IC, Linderman RG. Preinoculation of lettuce and onion with VA mycorrhizal fungi reduces deleterious effects of soil salinity. Plant and Soil. 2001;**233**:269-281

[113] Aroca R, Ruiz-Lozano JM, Zamarreño ÁM, Paz JA, García-Mina JM, Pozo MJ, et al. Arbuscular mycorrhizal symbiosis influences strigolactone production under salinity and alleviates salt stress in lettuce plants. Journal of Plant Physiology. 2013;**170**:47-55

[114] Tiwari S et al. *Pseudomonas putida* attunes morphophysiological, biochemical and molecular responses in *Cicer arietinum* L. during drought stress and recovery. Plant Physiology and Biochemistry. 2015;**99**:108-117

[115] Goswami D, Thakker JN, Dhandhukia PC. Simultaneous detection and quantification of indole-3-acetic acid (IAA) and indole-3-butyric acid (IBA) produced by rhizobacteria from l-tryptophan (Trp) using HPTLC. Journal of Microbiological Methods. 2015;**110**:7-14

[116] Naseem H, Bano A. Role of plant growth-promoting rhizobacteria and their exopolysaccharide in drought tolerance of maize. Journal of Plant Interactions. 2014;**9**:689-701

[117] Sandhya VZAS et al. Alleviation of drought stress effects in sunflower seedlings by the exopolysaccharides producing *Pseudomonas putida* strain GAP-P45. Biology and Fertility of Soils. 2009;**46**:17-26

[118] Bal HB et al. Isolation of ACC deaminase producing PGPR from rice rhizosphere and evaluating their plant growth promoting activity under salt stress. Plant and Soil. 2013;**366**:93-105

[119] Vardharajula S et al. Drought-tolerant plant growth promoting *Bacillus* spp.: Effect on growth osmolytes, and antioxidant status of maize under drought stress. Journal of Plant Interactions. 2011;**6**:1-14

[120] Ortiz N et al. The contribution of arbuscular mycorrhizal fungi and/or bacteria to enhancing plant drought tolerance under natural soil conditions: Effectiveness of autochthonous or allochthonous strains. Journal of Plant Physiology. 2015;**174**:87-96

[121] Shintu PV, Jayaram KM. Phosphate solubilising bacteria (*Bacillus polymyxa*)—An effective approach to mitigate drought in tomato (Lycopersicon esculentum mill). Tropical Plant Research. 2015;**2**:17-22

[122] Giri B, Kapoor R, Agarwal L, Mukerji KG. Pre-inoculation with arbuscular mycorrhizae helps *Acacia auriculiformis* grow in degraded Indian wasteland soil. Communications

in Soil Science and Plant Analysis. 2004;**35**:193-204

[123] Gaspar T, Penel C, Hadege D, Greppin H. Biochemical, molecular and physiological aspects of plant peroxidases. In: Lobarzewski J, Greppin H, Penel C, Gaspar T, editors. Plant Peroxidases. Geneva: Imprimerie Nationale, University of Geneva; 1991. pp. 249-280

[124] Jasper DA, Abbot LK, Robson AD. The effect of soil disturbance on vesicular-arbuscular mycorrhizal fungi in soils from different vegetation type. The New Phytologist. 1991;**118**:471-476

[125] Ruiz-Sanchez M, Aroca R, Monoz Y, et al. The arbuscular mycorrhiza symbiosis enhances the photosynthetic efficiency and the antioxidative response of rice plants subjected to drought stress. Journal of Plant Physiology. 2010;**167**:862-869

[126] Zarabi M, Alahdadi I, Akbari GA, Akbari GA. A study on the effects of different biofertilizer combinations on yield, its components and growth indices of corn (*Zea mays* L.) under drought stress condition. African Journal of Agricultural Research. 2011;**6**:681-685

[127] Lewis JA, Papavizas GC. Biocontrol of plant diseases: The approach for tomorrow. Crop Protection. 1991;**10**:95-105

[128] Chavoshi S, Nourmohamadi G, Madani H, Heidari H, Abad S, Alavi Fazel M. The effect of biofertilizers on physiological traits and biomass accumulation of red beans (*Phaseolus vulgaris* cv. Goli) under water stress. Iranian Journal of Plant Physiology. 2018;**8**:2555-2562

[129] Li Y, Shi H, Zhang H, Chen S. Amelioration of drought effects in wheat and cucumber by the combined application of super absorbent polymer and potential biofertilizer. PeerJ. 2018;7:e6073. DOI: 10.7717/peerj.6073

[130] Khosro M, Yousef S. Bacterial bio-fertilizers for sustainable crop production: A review. Journal of Agricultural and Biological Science. 2012;7:237-308

[131] Bhattacharyya P, Kumar R. Liquid biofertilizer-current knowledge and future prospect. In: National Seminar on Development and Use of Biofertilizers, Biopesticides and Organic Manures. West Bengal: Bidhan Krishi Viswavidyalaya Kalyani; 2000, November 10-12

[132] Verma M, Sharma S, Prasad R. Liquid biofertilizers: Advantages over carrier based biofertilizers for sustainable crop production. EnviroNews. 2011. Available from: https://isebindia.com/09-12/11-04-4

[133] Borkar SG. Microbes as Biofertilizers and their Production Technology. New Delhi, India: Wood Head Publishing India Pvt. Ltd.; 2015. pp. 7-153

[134] Mahdi SS, Hassan GI, Samoon SA, Rather HA, Dar SA, Zehra B. Bio-fertilizers in organic agriculture. Journal of Phytology. 2010;**2**:42-54

[135] Motghare H, Gauraha, R. Biofertilizers- types and their application. Krishi Sewa. 2012. Available from: http://www.krishisewa.com/articles/organicagriculture/115-biofertilizers.html

[136] Panda H. Handbook on Organic Farming and Processing. New Delhi: Asia Pacific Business Press Inc.; 2013. pp. 149-152

[137] Pathak AK, Christopher K. Study of socio-economic condition and constraints faced by the farmers in adoption of biofertilizer in Bhadohi district (Uttar Pradesh). Journal of

Pharmacognosy and Phytochemistry. 2019;**8**:1916-1917

[138] Mathur N, Singh J, Bohra S, Bohra A, Vyas A. Microbes as biofertilizers. In: Tripathi G, editor. Cellular and Biochemical Science. New Delhi: I. K. International Pvt. Ltd.; 2010. pp. 1089-1113

[139] Das D, Dwivedi BS, Meena MC, Singh VK, Tiwari KN. Integrated nutrient management for improving soil health and crop productivity. Indian Journal of Fertilisers. 2015;**11**:64-83

[140] Fierer N, Jackson RB. The diversity and biogeography of soil bacterial communities. Proceedings of the National Academy of Sciences. 2006;**103**:626-631

Chapter 4

Biochar: A Vital Source for Sustainable Agriculture

Kanayo Stephen Chukwuka, Akinlolu Olalekan Akanmu, Barachel Odaro-Junior Umukoro, Micheal Dare Asemoloye and Adegboyega Christopher Odebode

Abstract

The emerging concerns in sub-Saharan Africa are non-sustainability of agricultural and soil management practices threatening food security and environmental safety. Biochar, solid material obtained from thermochemical conversion of plants and/or animal biomass in an oxygen limited environment, is of great importance both agriculturally and environmentally. This chapter reviews the contributions of "biochar technology" to environmental sustainability and food security. This strategy addresses the declining food security issues, depleting soil and plant health challenges. When properly exploited, biochar will enhance soil fertility recovery, guarantee resilience to climate change challenges, and satisfy food production needs of growing global population. The positive impacts of biochar utilization on soil beneficial organisms in harnessing and controlling pests and diseases as well as revitalization of ecological niche make it a preferred option. Unfortunately, there is dearth of information on biochar mechanism to enhance bioremediation technology, which is still facing some challenges that need attention for adequate soil remediation. Many researchers have demonstrated bioremediation in laboratory scale under controlled environmental conditions; it may however be very problematic to establish the growth/survival of these biological entities in situ on heavily polluted soil where the environmental conditions cannot be controlled.

Keywords: biochar, plant productivity, environmental safety, bioremediation, food security

1. Introduction

Food security and environmental safety are the emerging concern in sub-Saharan Africa due to non-sustainable agricultural and soil management practices [1]. Thus, giving rise to the limiting influence of biotic and abiotic stress factors on the plant and soil health [2]. Asides the resulting declined in agricultural production, the contributory effect of soil contamination by industrial pollution and excessive use of chemical in agriculture presently constitute a threat to food security and environmental safety. Therefore, this review examined the prospects of biochar in the sustainable agricultural production, plant protection and soil restoration. Biochar is a solid waste material obtained from thermochemical conversion of plant or animal biomass or both in an oxygen limited environment [3].

The thermal process is carried out on sources of biomass including agricultural wastes, green-waste, and animal manures (biomass feedstock) at temperatures ranging from 200 to 900°C [4, 5]. Biochar addition to soils was engineered by Amazonian terra preta soils, which were characterized by high levels of fertility as compared to adjacent soils where no organic carbon addition occurred [6]. The overall benefit of biochar to environment and production systems is based on three sustainability factors; use of sustainable biomass, sustainable production processes, and sustainable end-use [7]. While the main applications which prompted its recent scientific research include: its mitigation of climatic change, efficient and cost effective waste management, and the use of biochar as amendment to improve soil quality and sustain crop yield [8, 9]. Therefore, biochar amended soil causes alterations in soil health and this encompasses physical, chemical and biological features while it as well maintains the functions of both natural and managed ecosystems essential for sustainable agricultural fertility and productivity [10, 11].

2. Interaction of biochar with soil, plant, and microorganisms

2.1 Interactions between biochar and soil

Biochar exhibits natural oxidation through the formation of functional groups, thereby providing sites that can retain nutrients and other organic compounds [12]. Through the association of biochar particles with clay and silt-sized minerals, oxidized biochar particles may be bound to soil minerals, thereby decreasing the potential of its decomposition [13]. Hence, enhancing the ability of soil biochar complex to adsorb organic compounds present in the soil while biochar also interact directly with organic matter of soil by sorption [14]. Generally, soil health are restored with amendments by balancing its pH, increasing organic matter content and water holding capacity, re-establishing microbial communities, alleviating compaction and structure thereby allowing establishment of vegetation, recreate ecological function of soils, decrease bioavailability of toxic pollutants, leachability and mobility of contaminants, erosion, improve soil drainage and reduce costs compared to traditional remediation techniques [15]. The cations in biochar after pyrolysis are transformed into oxides, hydroxides, and carbonates (ash) which act as liming agents when applied to soil. Biochar is composed of low density material that reduces soil bulk density, thereby increasing water infiltration, root penetration, soil aeration and aggregate stability [16]. Biochar amendments on soil have positive effects on nutrient retention, particularly in highly weathered soils with low ion-retention capacities [17]. Biochar application to medium and coarse textured soils increased soil water holding capacity when analyzed [18]. Thus, biochar serves as soil amendment and carbon sequestration medium [19].

2.2 Impact of biochar on nitrogen fixation

Biochar as soil amendment enhances the biological nitrogen fixation, the nitrogen available in soil is usually lower than that of the biochar due to the high carbon/nitrogen (C/N) ratio of the biochar, and the resulting N immobilization [17, 19], as well, contains higher availability of nutrients, N, and almost neutral pH value [20]. Combination of factors related to soil nutrient availability and simulation of plant microbe interaction, along with nitrogen/nutrient levels also increases when biochar was applied to soil resulting in increased colonization of the host plant roots by *Arbuscular mycorrhiza* fungi (AMF). Biochar amended soils enhanced biological N-fixation in leguminous crops as reported by Rondon et al. [21]. The increase in the availability of major plant nutrients due to application of biochar occurs as the biochar also releases some small amounts of nutrients that would be available to soil biota [22].

2.3 Effect of biochar application on plant productivity

The prevailing scientific understanding of biochar degradation in soils is that some portions of it are quite readily decomposable, while the core structure of the material is highly resistant to degradation. However, biochar promotes plant productivity and yield through several mechanisms, it changes the physical conditions of plants; its dark color alters thermal dynamics and facilitates rapid germination, allowing more time for growth compared with biochar un-amended soil [23]. Although, there are no specific recommended application rates for any soil but amendment of soils must be done based on extensive field testing. Chan et al. [24] reported that application of 5–50 tons of biochar per hectare, with appropriate nutrient management had positive effects on crop yields. Single application of biochar can provide beneficial effects over several growing seasons in a field due to its recalcitrance to decomposition in soil [25]. Therefore, biochar does not need to be applied in all cropping season, as is usually the case for manures, compost, and synthetic fertilizers. It effects on yield also occur as a result of changes in soil nutrition, water holding capacity and microbial activity and these effects vary due to soil type [26].

Many researcher has affirm the importance of biochar in plant growth enhancement, hardwood biochars and poultry manure biochars possesses nutrients, such as, high N content, which often enhance positive yield increases [26]. Petter et al. [27] reported that *Eucalyptus* biochar positively affected upland rice yields since the first year of its application. Biochar produced from wood, paper pulp, wood chips and poultry litter have also been found to positively affect crop and biomass yield [18]. In the studies of Glaser et al. [17] and Chan et al. [24], corn, cowpea and radishes grown on poultry litter biochar, each of their yield was improved by 140, 100 and 96% respectively. Field application of biochar below 30 tons/ha was reported to increase crop productivity and varied with crop type with greater increases for legume crops (30%), vegetables (29%) and grasses (14%) compared to cereals, such as, corn (8%), wheat (11%) and rice (7%) [28, 29]. Furthermore, wastewater sludge biochar was applied to cherry tomatoes at the rate of 10 tons/ha and resulted in 64% increased production above the control soil conditions [30].

Combined application of pine woodchip biochar at the rates of 5 and 10 mg/ha with N fertilizer to a fertile silt loam soil in northwest Arkansas significantly increased corn yield compared to sole application of N fertilizer [30]. Combined application of biochar with cow urine to the root zone of pumpkin also significantly increased pumpkin yield compared to all other soil amendment treatments. Ndor et al. [31] reported that rice husk and sawdust biochar significantly increased N, P and K uptake by maize plant, and also significantly increased maize number of leaves, plant height, fresh and dry weight of cobs. In another study, amendment of an alkaline soil with biochar derived from vegetable waste and *Eucalyptus*-leaves had significant effects on seedlings dry matter, shoot and root lengths of maize [32]. According to Fru et al. [33], *Talinum triangulare* responded positively in growth, nutrient uptake and yield when cultivated in poor and acidic soil amended with biochar. Instances of decreasing yield due to a high biochar application rate were reported when equivalents of 165 tons of biochar/ha was added to a poor soil in a

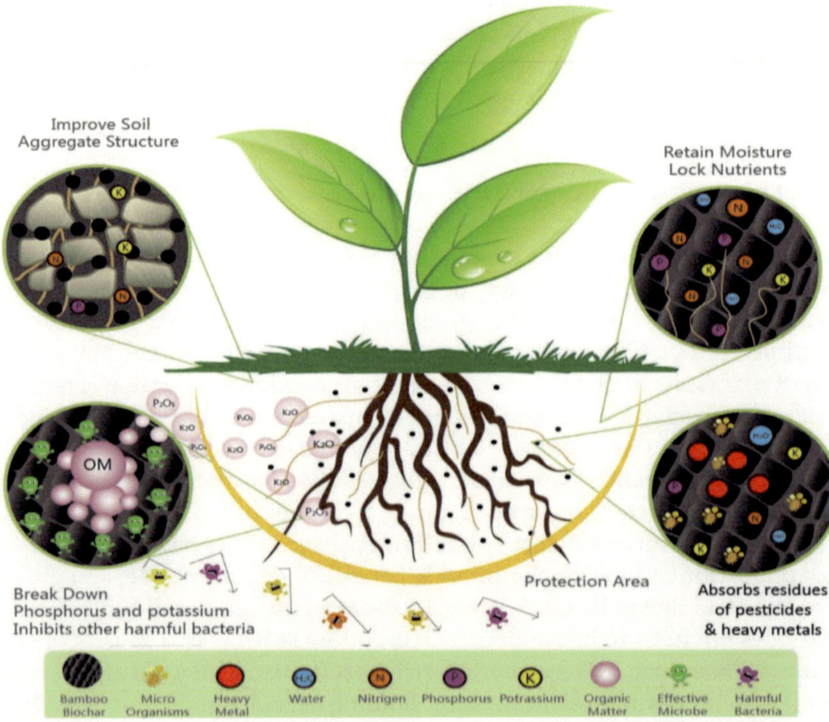

Figure 1.
Effect of biochar on the soil mineral component. Source: Bamboo biochar—Bio-organic fertilizer [35].

pot experiment [21]. Yield increase was reported in maize and wheat when biochar was combined with either organic residue/compost or mineral fertilizer, this indicates that wood biochar may raise nutrient use efficiency when added to organic/inorganic fertilizer/crop residues [34] (**Figure 1**).

2.4 Effects of combined application of biochar and fertilizer

Application of biochar in combination with fertilizer to soil has been found to have positive effects on crop growth, this was probably due to the positive interaction between biochar and applied fertilizer that improved the availability of nutrients associated with enhanced plant uptake and reduced losses of these nutrients [25, 35, 36]. Most biochar materials are not substitutes for fertilizer, so adding biochar without necessary amounts of nitrogen (N) and other nutrients cannot be expected to provide improvements to crop yield [21]. Application of *Eupatorium* (Syn. *Chromolaena*) weed-derived biochar to soil increased the yield of pumpkin crop to 85% above the unamend soil. A normal application rate in the range of 5–20 t/ha of biochar similar to other amendments, such as, compost under normal conditions can positively affect crop yield while excessive application rates (>50 t/ha) may negatively affect crop response. However it will be difficult to establish an exact threshold above which negative effects appear [37]. Negative effects on crop growth are mostly reported with biochar obtained from municipal waste, food waste, and sewage sludge because their excessive sodium contents increase soil salinity [28]. Other negative effects from plant- and wood-based biochars are due to one of the following causes: high application rates, high volatile matter contents detrimental to crop growth, reduced plant available nitrogen, or negative liming effect in alkaline and calcareous soils [35].

3. Role of biochar in sustainable plant disease management

3.1 Plant disease management: the good and the bad

Due to the increasing global population, there is an ever increasing desire to increase agricultural efficiency in terms of producing maximum crop yields and produce. This is only achievable if pest and disease agents limiting crop productions are adequately checked. Cultural, biological, chemical and regulatory measures are the key methods of plant disease management. Since its introduction over a century ago, chemical method had assumed a position of importance and preferred over the existing cultural method due to its effectiveness in the control of diseases, pests or weeds. The relatively low cost of the chemicals, the ease with which they can be applied, availability, stability and fast-acting limits the damage done to crops. However, with the realization of the havoc caused by continuous and persistent use of chemicals either by misuse or abuse, with the consequent degradation of ecological community of most of the farm sites based on their effects on both the target and non-target organisms, has led to the destruction of beneficial organisms and the natural predator in the eco-system. They also obstruct the normal functioning of the ecosystem if the pests and organisms' develop resistance to the chemicals used, thus resulting in pests evolution. However, agricultural workers often suffer occupational exposure to pesticides while exposure of the entire population is exposed to pesticides pollution primarily through the food chain and drinking water contaminated with pesticide residues which are carcinogenic [38, 39].

3.2 Biological control: novel strategy to safe agricultural practices

Humanity does not only dependent on the direct contributions of microbes within our bodies but also on the way they shape and maintain essential functions of our environment, including agricultural production systems where they provide "ecosystem services". Therefore, the use of biological control in the management of pest and diseases pre-dates the modern pesticide era [40]. The host specificity, longer residual effect and non-toxicity to human and the environment makes biological control a novel strategy to safe agricultural practices and sustenance of the ecosystem structure [40]. Some of the control measures that have been widely explored in the management of plant pathogens include the use of beneficial organisms [41]. These are mostly members of the bacterial genera *Bacillus*, *Pseudomonas*, *Serratia*, *Stenotrophomonas*, and *Streptomyces* and the fungal genera *Ampelomyces*, *Coniothyrium*, and *Trichoderma* being used as the model organisms to demonstrate their influence on plant health [42]. The growth and health enhancement of *Arbuscular mycorrhizal* fungi (AMF) has been well investigated [43–45] while the use of plants which involves its extracts, metabolites and bioactive products had been widely explored and yielded positive responses in the management of phytopathogens of varying kind of agricultural plants [46–48].

3.3 Influence of biochar on soil biota

Soil biota is important to the functioning of soils and provides many essential ecosystem services. According to Wuddivira et al. [49], biochar amended soil provides suitable pH for the growth of microbes, especially fungal hyphae due to its porosity. Application of biochar into soils leads to initial degradation of biochar by chemical oxidation and microbial processes [50]. These processes that influence the energy flow and organic matter within the soil will impinge on bacterial and fungal-based energy channels, which will have impact at higher trophic levels [51]. Biochar and earthworms increased the availability of mineral nutrients in growing seedlings suggesting that this mechanism played important role [52]. Microbial population could be higher in black carbon rich soil, thus the interaction between biochar as a soil amendment approach plays a vital role in soil biota [53]. More so, biochar amendment resulted in increased soil microbial biomass and changes in the composition of soil microbial community [54].

3.4 Biochar: the refuge for microorganisms

Biochar provides "sanctuary" for microorganisms experiencing harsh environmental conditions and also provide a more habitable space for their proliferation. The types of biochar and soil environment are factors that are vital to how the char will affect the soil microbiota [41, 55]. The ability of biochar to act as refuge for soil microorganisms from their predators; protozoa, beneficial nematodes, and microarthropods was affirmed in the report of Verheijen et al. [56] which showed the electron microscopic images of bacteria and fungi on the surface or inside biochar pores. The larger biochar pores which are mostly the structural remnants of wood xylem, phloem vessels and other larger features have diameters >10 μm [57]. According to IUPAC conventions, biochar porosity has been classified by distinguishing between; micropores (<2 nm), mesopores (2–50 nm), and macropores (>50 nm) [58, 59]. Whereas, the organisms that predate soil bacteria, including protozoa and nematodes, have diameters <10 μm therefore making access to the larger pores relatively easy [57]. Further protection from predation offered to the microbes by biochar has been associated with the hydrophobic adsorption of biochar although the hydrophobicity can decrease over time. Some microorganisms can be strongly attached to hydrophobic surfaces which create biofilms of several bacterial layers thick [57, 59].

3.5 Role of biochar in environmental safety and sustainable agriculture

The benefits of biochar on crop productivity and plant health have been related to four main mechanisms which include; increase in soil pH which is beneficial to acidic soils [18]. Biochar's high water retention capacity results to improvement of water regime of the soil, this is of special advantage to sandy soil area where biochar will reduce the leaching away of moisture, thereby reducing water loss, while it reduces the risk of water-logging in clay soil by promoting water drainage [60, 61]. The third mechanism is associated with the capability of biochar to adsorb and neutralize phytotoxic organic molecules including anthropogenic, xenobiotics and natural allelopathic compounds. This detoxifying capability is directly related to the dramatic increases of specific surface area that occur during pyrolysis [62–65]. The fourth mechanism is related to its capability to stimulate beneficial microbes, in bulk soil as well as in the rhizosphere [66]. By serving as a source of reduced carbon compounds and by increasing the availability of micronutrients, biochar may be beneficial to microbial populations, such as, *Arbuscular mycorrhizal* fungi (AMF) [19, 67], plant-growth-promoting microbes [68, 69]. Therefore, biochar applications increase the microbial biomass of the beneficial organisms with related changes in microbial community functionality [66, 69]. However, the increase in microbial biomass resulting from microbial growth following biochar application has been reported to be as a result of the; effect of water and nutrient retention, formation of active surfaces that provided optimal habitat for microorganisms, weak alkalinity and partial inhibition of destructive and simultaneous support for beneficial microorganisms [70, 71].

3.6 Biochar-microbe interaction: mode of action in plant disease control

Treatments with resistance inducers or beneficial microorganisms have been reported to provide long-lasting resistance for plants to a wide range of pathogens [72]. More so, induced resistance can be also conferred by plant-associated microorganisms, including beneficial bacteria and/or fungi [41]. Biochar does not have an indigenous population of microorganisms that can potentiate disease suppression, due to the high thermal treatment in its production [7]. However, its addition influences microbial populations and communities, thus causing changes which may include increase in beneficial microorganisms that directly protect plants against soil pathogens by; producing antibiotics, out-competing the pathogens, or grazing on the pathogens [7, 9, 73]. Investigations conducted on biochar and microbe interaction collated by Bonanomi et al. [9] proposed five different mechanisms by which biochar mitigate against plant diseases and these include: (i) induction of systemic resistance in the host plants; (ii) enhanced abundance and/or activities of beneficial microbes; (iii) modification of soil quality in terms of nutrient availability and abiotic conditions; (iv) direct fungitoxic effects of biochar; (v) sorption of allelopathic and phytotoxic compounds. This attributes have been further verified in some of the recent investigations [19, 72, 74, 75], thus biochar is an evolving strategy and a potent tool in plant pathology research, **Figure 2**. Therefore, biochar is gaining importance day by day as its application touches all facet of agriculture and has attracted tremendous attention in the practice of sustainable agriculture.

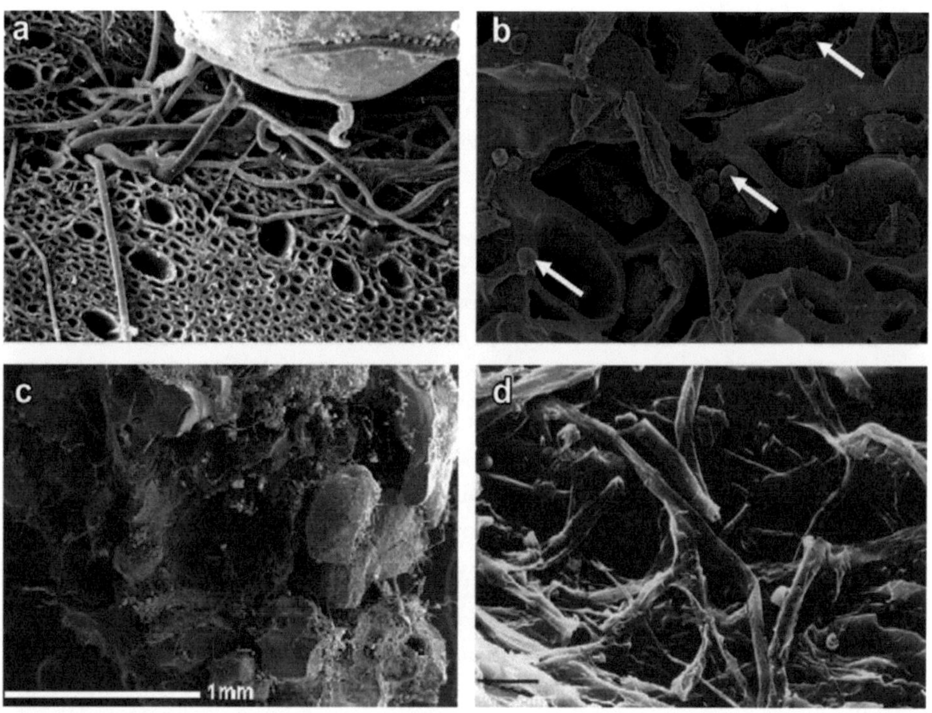

Figure 2.
Spatial association and colonization of biochar by microorganisms (a) fresh biochar showing fungal hyphae; (b) fresh corn biochar showing microorganisms in pores (arrows); (c) 100-year-old char from a forest fire isolated from a frigid entic Haplorthod; (d) 350-year-old char from a forest fire in a boreal forest soil. Source: Lehmann et al. [19].

3.7 Biochar interaction with mycorrhizae

Biochar and *Arbuscular mycorrhizal* fungi (AMF) interaction in soil will alter levels of nutrient availability that affects both plants and mycorrhizal fungi communities and modifies plant-mycorrhizal fungi complex which serves as a refuge from hyphal grazers and soil predators [76]. Biochar soil amelioration in degraded landscapes has the potential to increase grassland plant production, enrich soil microbial populations, and stimulate *Arbuscular mycorrhizal* persistence while addition of biochar to soil increases root colonization by AMF [22]. In addition to the mineral supplement of AMF by biochar, it also acts against biotic and abiotic stresses in nature thus increasing the ability of AMF to assist their host in resisting infection by plant pathogens [73].

4. Effects of biochar application on greenhouse gases

Global surface temperature has increased by 0.8°C in the last century primarily because of increased anthropogenic emissions of long-lived greenhouse gases (GHGs), such as, carbon (iv) oxide (CO_2), nitrous oxide (N_2O) and methane (CH_4). Greenhouse gases are those that adsorb and emit radiation within the thermal infrared range [77]. Application of biochar to soils can impact soil GHG fluxes by changing the composition and activity of soil microbes, soil pH and soil biogeochemical processes [24]. Effects of biochar amendment on soil GHG fluxes depend on the study conditions, duration of the experiment, biochar application rate, biochar feedstock and pyrolysis methods [78].

4.1 Effects of biochar application on soil CO_2 emissions

Soil CO_2 emissions can be derived from native soil organic matter, the mineralization of added carbon compounds (such as, dead plant material), root exudates or dead roots and the direct respiration from plant roots [79]. Lehmann et al. [19] suggested that a co-benefit of biochar amendment is a reduction of soil CO_2 emissions and associated long-term increases in soil organic carbon (SOC) in the soil. Although the mechanisms governing the effects of biochar amendment on soil CO_2 emission are uncertain, some authors suggested that increased CO_2 emissions from soil might be as a result of the following mechanisms: biochar reduces the albedo of the soil increasing soil temperature; addition of liable carbon, increased substrate for soil carbon mineralizing enzymes [80]; agglomeration of soil carbon, microbes, nutrients on biochar surface and increased carbon use efficiency; reduction of carbon mineralizing enzymes activity and soil-derived CO_2 precipitation onto the biochar surface as carbonates [81]. Wang et al. [82] reported that addition of wheat derived-biochar to acidic soil increased soil organic C and CO_2 efflux on average by 61 and 19%, respectively.

4.2 Effects of biochar application on soil N_2O emissions

Nitrous oxide is produced in soils primarily by microbial activity through nitrification, nitrifier denitrification, NO_3 ammonification and denitrification [83]. Biochar amendment to soil can have significant effects on soil N_2O emissions; however, the magnitude of effect varies widely. According to Yanai et al. [84] and Stewart et al. [85] short-term laboratory incubations have shown that biochar amendment can suppress soil N_2O emissions, while Spokas [86] and Jones et al. [87] concluded that soil N_2O emissions were not suppressed with biochar amendment in the longer term (up to 3 years after biochar addition). Biochar amendment causes changes to a range of soil physical and chemical properties that regulate N-cycling processes [86]. Some authors have explained the mechanisms of the effects of biochar amendment on soil N_2O emission and these include: increased water holding capacity and decreased bulk density of the soil, increased soil aeration thereby reducing the activity of denitrifying microorganisms [84, 88], reduced N substrate for nitrifying and denitrifying enzymes thereby reducing enzymatic activities of soil microbes as a result of immobilization of soil inorganic N through absorption to biochar surface or increasing microbial immobilization [86], the N_2O:N_2 emission ratio produced during denitrification decreases as a result of increased soil pH [89]; N_2O:N_2 product ratio of denitrification is reduced by increased effects to the soil [90], reduced activity of soil nitrifying/denitrifying organisms through substances emitted by the biochar, such as, ethylene, α-pinene, PAHs, VOCs [91]. Rondon et al. [92] reported that biochar amendment reduced N_2O emissions from pasture land and soybean soil by 80 and 50% respectively, because microbial conversion and denitrification were restricted. Application of biochar at a rate of 40 t/ha decreased N_2O emission from paddy rice and maize fields by 21–28 and 10.7–41.8%, respectively but increased CH_4 emission from a paddy rice field by 41% and CO_2 emission from a maize field by 12% [93, 94]. Cayuela et al. [95] reported that biochar reduced soil N_2O emissions by 28% in a similar field. Wang et al. [82] reported that addition of wheat derived biochar to acidic soil did not affect the annual N_2O emissions (26–28 kg N/ha), but reduced seasonal N_2O emissions during the cold period. Fan et al. [96] reported that biochar amendments generally stimulated the NH_3 emissions with greater enhancement from wheat straw biochar than swine manure biochar.

4.3 Effects of biochar application on soil methane (CH_4) emissions

Methane are produced by methanogens as a metabolic by-product of organic matter mineralization in anaerobic conditions; the two primary pathways being through CO_2 reduction by H_2 or through acetotrophy [97]. Soil methanotrophs are the only known biological sink for atmospheric methane, which oxidize methane and produce CO_2 as a by-product [98]. Soil methanotrophs require oxygen as a terminal electron acceptor and their activity is highest around 60% water-filled pore space (WFPS) and decreases above this moisture content [88, 99]. Zhang et al. [93] and Wang et al. [100] reported that there are limited evidence to suggest that biochar amendment affects soil CH_4 emissions, and evidence that supports it are mostly from studies in rice paddies. In saturated soils, such as, rice paddies but not in other aerobic crop soils, CH_4 emissions are generally significant [97]. Increased availability of liable C substrates for methanogenic bacteria may explain increased methane emissions following the addition of biochar to soil [100]. Soil CH_4 emissions were increased by 37% with biochar amendment in a paddy rice field [100]. Similar observation of an increase in soil CH_4 emission from the same land use were also reported by Zhang et al. [93] and Knoblauch et al. [101]. Increased in soil methane uptake within arable soils following biochar amendment was observed by Karhu et al. [88]. In similar studies, with other crop types, no significant effect of biochar amendment on CH_4 emissions in arable and pasture soils were reported [100, 102, 103] while in Finnish agricultural soil, a 96% increase in methane uptake was reported in biochar amended soils [88]. Application of biochar to waterlogged paddy rice soil in the laboratory, decreased CH_4 and CO_2 emissions in the soil and this was attributed to the restriction in methanogen activity and limitation of carbon on microbial biomass, as well as rise of pH value [104]. However, increase in CH_4 and CO_2 emissions were reported by Ameloot et al [105] and Van Zwieten et al. [106]

studying the short-term CO_2 and N_2O emissions and microbial properties of biochar amended sandy loam soils, and effects of biochar from slow pyrolysis of papermill waste on agronomic performance and soil fertility, respectively.

5. Characterization of biochar as an effective mediator of bioremediation mechanisms

5.1 Biochar as a soil additive to enhance soil restoration/remediation

Soil remediation efforts should be based on feasible, environment friendly and cost effective technologies, and many scientists today are advocating bioremediation mechanisms for meeting these criteria. It has also been widely reported that bioremediation can be enhanced through the use of traditional resources, such as, the application of soil additives. The basis for the use of soil additives during bioremediation is for enhanced bio-stimulation and bio-augmentation; these two mechanisms form the bed-rock for the immense roles of soil additives in bioremediation as reported by many researchers [45, 107–112]. Biochar applications as a soil additive in contaminated soil is a potential management strategy for feasible and cost effective agricultural sustainability using degraded soils hence improving the food security. Many reports have identified biochar with high sorption capacity for many contaminants including persistent organic pollutants (POPs) and many inorganic pollutants, such as, heavy metals [113, 114]. It was reported, however, that the physicochemical properties of the original crop residue used for biochar preparation may determine its sorption efficiency [115, 116]. However, good knowledge of the pollutant type and its concentration may help in predicting the type of biochar that would be of best fit. Thus it is a crucial factor to clarify the correlation between the sorption efficiency and properties of a particulate. The use of biochar in nutrient sequestration according to Barrow [117] was from the discovery of "terra preta" which is a charcoal-rich fertile soil located at the central Amazon basin that is known for diverse agricultural roles. The importance of such biochar to soil has been reported [28, 105, 117–122].

In a study by Gomez-Eyles et al., biochar was shown to reduce PAH accumulation in earthworm (*Eisenia fetida*) tissue; this organism was incubated in soil treated with biochar for 28 and 56 days. Their study suggested that biochar can be used in PAH polluted soil to avoid their entrance into the food chain and this was corroborated in another study reported by Wang et al. [110] in which the bioavailability of pesticide called chlorantraniliprole was reduced by biochar amendment and prevented its absorption in earthworm tissues. The use of biochar as amendment for enhanced bioremediation is gaining attention at an exponential rate as it enhances soil nutrients and water availability [115–117]. It was also reported that it functions by immobilizing/degrading many soil and water contaminants [123, 124]. Unlike other amendments, biochar is thought to be perfect in Carbon sequestration in soil making it stable for several years [122].

5.2 Biochar as a potential catalyst for phytoremediation

The use of plants for bioremediation is called "phytoremediation"; its success depends on establishment and good development of vegetation on the polluted site brought about by healthy root and shoot biomass [125, 126]. However, the major problem of phytoremediation is the establishment of degree/level of pollution of the polluted site. Many soil pollutants are very persistent, many form complexes with humus and soil nutrients and thus make them unavailable for plant use. In addition early research have reported that many soil pollutants, such as, aromatic and polycyclic hydrocarbons can create anaerobic condition and make seed germination and plant establishment difficult on the polluted soil. However, the use of amendments, such as, organic materials can enhance plant biomass yield and improved plant health growing on polluted sites [1, 111, 127]. The use of biochar for the amendment of polluted soils has been reported to enhance re-establishment of plants and supports massive plant biomass [125, 128]; this is a potential approach for effective phytoremediation mechanism [128]. Prendergast-Miller et al. [129] and Prendergast-Miller et al. [130]. The above reports present biochar as good candidate for positive soil enhancement although this is not yet fully exploited in contaminated sites. Biochar as soil amendment enhances nutrient availability and improves the activities of soil microorganisms around the root rhizosphere for effective nutrient mobilization for root uptake [116, 131–133]. This promotes root expansion and hence has potential to support phytoremediation. It was reported that plant growing on polluted soil may develop some disease malformations due to their response to the toxic pollutants [111] and reduced resistance to pests.

Biochar has however been reported to increase plant resistance to several soil and air-borne pathogen, it was reported that biochar stimulates several plant's defense pathways and related gene expression [7, 134, 135]. Biochar has also been well reported for its pH enhancement capacity as it enhances the soil CEC [136–139] and minimizes salt toxicity in polluted soils (142). According to Cheng and Lehmann [140] and Singh et al. [115], biochar is able to oxidize in soil to raise its carbon exchange capacity (CEC) as it disintegrates during tilling and weathering and there are some commercially available biochar purposefully used as soil nutrient enhancement. This can be very helpful in the enhancement of plants health during phytoremediation. Many researches have suggested the combination of phytoremediation with biochar for effective soil remediation. Hartley et al. [141] and Fellet et al. [142] reported that biochar's combined use with *Miscanthus* increased phytostabilization, while in another experiment, Hartley et al. [141] reported an increase in As extracted by *Miscanthus* plant in three soils treated with biochar made from hardwood. Combined biochar and phytoremediation have also been reported in Cd-polluted soils using *Brassica napus* L. by Houben et al. [143]. Many more results have proved that biochar is suitable for enhancement of phytoremediation mechanism and this seems plausible for their exploitation in remediation of multi-contaminated soils. Biochar can be used prior to plant colonization of acidic soils however; these two approaches still need more confirmatory researches to depict their synergistic mechanisms for effective and sustainable set-up.

5.3 Biochar as a potential catalyst for enhancement of microbial response and bioremediation

The nutrient and soil amendment capability of biochar cannot be overemphasized. The fact that biochar improves soil nutrients means that it has some beneficial effects on the soil microbiota, this has been well studied [115], Ippolito et al. [144, 145] and Kuppusamy et al. [146] and they reported that biochar amendment was found to enhance increment in microbial biomass, diversity and enzyme activities [14, 147, 148]. Biochar was thought to increase bacterial and fungal population as they easily form pore habitats in biochar, although the mechanism could be varied but the overall effect might be as a result of nutrient recycling and soil water retention as enhanced by the biochar and this of course increases the resources available for the microbes to use.

Furthermore, microbial community shifts brought about by biochar amendments may vary due to particular characteristics, microbial response to soil conditions set by the biochar treatments, especially based on the biochar surface characteristics and its bioavailable compounds, as well as well as pH changes induced by the treatment [149]. Zimmerman et al. [80], for example, reported an increase in C mineralization in soil treated with biochar that was made from grasses under low temperatures of 250–400° C compared to biochar that was made from hard woods under higher temperatures of 525–600°C. In another study, Steinbeiss et al. [150] explained that fungi adapted more with biochar that was created from yeast but Gram-negative bacteria responds well to those created from glucose. Anderson et al. [151] reported varying results in microbial population dynamics due to different biochar treatments. The exploitation of biochar to increase microbial population for faster bioremediation requires complete understanding of the relationship of a particular biochar to be used and the type of native or introduced bacteria or fungi involved. Theoretically, biochar is not a degrading substrate and therefore cannot be degraded by microbes [149, 152]; however, it has a potent labile C source in itself [153]. When used in soil, biochar adsorbs active enzymes and nutrients and make them available for usage by microorganisms [154].

5.4 Prospects of biochar as a potential mediator for synergistic bioremediation mechanisms

Interest in combining different biodegradation mechanisms, such as, the use of plant and microbial communities for effective synergistic bioremediation has been vogue in some years past. It is believed that bioremediation which provides an environmentally friendly mechanism has gained acceptance by scholars and environmental managers. Biochar is a synergistic bioremediation mechanism known to yield speedy soil remediation if correctly implemented [155]. The synergistic combination of two or more biological entities for effective soil bioremediation requires that the two or more organisms to be combined must integrate well and can coexist together leading to exchange of mutual benefits between one another. This requirement is paramount in such settings as bioremediation such that many researchers have suggested the use of soil supplements/additives to enhance this technology [112, 155]. Having established the importance of biochar for plant and microbes, the smart way therefore is to think about using it for combine plant-microbial bioremediation technology. Nutrient enhancement by biochar fosters root elongation and hence increases the microbial populations in the rhizospheres [18, 156]. It is anticipated that effective combinations of different bioremediation technologies may eventually yield a feasible, speedy and effective means of restoration of many polluted soils, and basic roles of biochar in this area cannot be overlook. However, this requires a good understanding and interest in biochar properties, mode of production and their actions in different polluted environments need to be well studied. Large and small scale laboratory and field trials are needed for proper exploitation of this technology.

6. Conclusion

Considering the arrays of biochar benefits, it is a potentially untapped asset for sustainable soil health. However, the dearth of adequate research and knowledge of biochar use as soil amendment today is a still big gap. This therefore necessitates mechanistic understanding and research to unveil the mechanisms of biochar action on soil health. It would as well enhance knowledge on the optimal rate for a particular biochar application, its quality parameters/suitability to different soil and climatic conditions, in relation to economic factors, feedstock properties and optimize designed pyrolysis conditions needed for its production toward specific end use. In addition, most of the studies on biochar so far are generally based on short term; long-term experiments are needed to understand the effect of biochar aging. Having fulfilled these recommendations, biochar may well be one of the prominent scientific breakthroughs for benefit of mankind.

Author details

Kanayo Stephen Chukwuka[1], Akinlolu Olalekan Akanmu[1*], Barachel Odaro-Junior Umukoro[1], Micheal Dare Asemoloye[1,2] and Adegboyega Christopher Odebode[1]

1 Department of Botany, University of Ibadan, Ibadan, Oyo State, Nigeria

2 Health Science Platform, School of Pharmaceutical Science and Technology, Tianjin University, P.R. China

*Address all correspondence to: akinakanmu@gmail.com

References

[1] Chukwuka KS, Omotayo OE. Soil fertility restoration potentials of Tithonia green manure and water hyacinth compost on a nutrient depleted soil in South Western Nigeria using *Zea mays* L. as test crop. Research Journal of Soil Biology. 2009;**1**(1):20-30

[2] Atkinson NJ, Urwin PE. The interaction of plant biotic and abiotic stresses: From genes to the field. Journal of Experimental Botany. 2012;**63**(10):3523-3543

[3] International Biochar Initiative. Standardized product definition and product testing guidelines for biochar that is used in soil. IBI biochar standards. 2015. Available from: https://www.biochar-international.org/wp-content/uploads/2018/04/IBI_Biochar_Standards_V2.1_Final.pdf [Last accessed: 21-05-2019]

[4] Yuan JH, Xu RK. The amelioration effects of low temperature biochar generated from nine crop residues on an acidic Ultisol. Soil Use and Management. 2011;**27**:110-115

[5] Conversa G, Bonasia A, Lazzizera C, Elia A. Influence of biochar, mycorrhizal inoculation, and fertilizer rate on growth and flowering of *Pelargonium* (*Pelargonium zonale* L.) plants. Frontiers in Plant Science. 2015;**6**:429

[6] Harder B. Smoldered-earth policy: Created by ancient Amazonian natives, fertile, dark soils retain abundant carbon. Science News. 2006;**169**:133

[7] Elad Y, Cytryn E, Harel YM, Lew B, Graber ER. The biochar effect: Plant resistance to biotic stresses. Phytopathologia Mediterranea. 2011;**50**:335-349

[8] Sohi SP, Krull E, Lopez-Capel E, Bol R. A review of biochar and its use and function in soil. Advances in Agronomy. 2010;**105**:47-82

[9] Bonanomi G, Ippolito F, Scala F. A "black" future for plant pathology? Biochar as a new soil amendment for controlling plant diseases. Journal of Plant Pathology. 2015;**97**(2):223-234

[10] Chintala R, Owen R, Kumar S, Schumacher TE, Malo D. Biochar impacts on denitrification under different soil water contents. World Congress of Soil Science. 2014;**6**:157-167

[11] Kumar S, Nakajima T, Mbonimpa EG, Gautam S, Somireddy UR, Kadono A, et al. Long-term tillage and drainage influences on soil organic carbon dynamics, aggregate stability, and carbon yield. Soil Science & Plant Nutrition. 2014;**60**:108-118

[12] Cheng CH, Lehmann J, Engelhard MH. Natural oxidation of black carbon in soils: Changes in molecular form and surface charge along a climosequence. Geochimica et Cosmochimica Acta. 2008;**72**:1598-1610

[13] Brodowski S, John B, Flessa H, Amelung W. Aggregate-occluded black carbon in soil. European Journal of Soil Science. 2006;**57**:539-546

[14] Hammes K, Schmidt WI. Changes of biochar in soil. In: Lehmann J, Joseph S, editors. Biochar for Environmental Management: Science and Technology. London: Earthscan; 2009. pp. 169-182

[15] U.S. EPA. The Use of Soil Amendments for Remediation, Revitalization and Reuse. Hazardous Waste Clean-up Information System (Clu-In). 2007. Available from: www.Clu-in.org/pub1.cfm

[16] Laird DA, Fleming PD, Davis DD, Horton R, Wang B, Karlen DL. Impact of biochar amendments on the quality of a typical Midwestern agricultural soil. Geoderma. 2010;**158**:443-449

[17] Glaser B, Lehmann J, Zech W. Ameliorating physical and chemical properties of highly weathered soils in the tropics with charcoal—A review. Biology and Fertility of Soils. 2002;**35**(4):219-230

[18] Jeffery S, Verheijen FGA, Van Der Velde M, Bastos AC. A quantitative review of the effects of biochar application to soils on crop productivity using meta-analysis. Agriculture, Ecosystems and Environment. 2011;**144**:175-187

[19] Lehmann J, Rillig MC, Thies J, Masiello CA, Hockaday WC, Crowley D. Biochar effects on soil biota—A review. Soil Biology and Biochemistry. 2011;**43**:1812-1836. DOI: 10.1016/j.soilbio.2011.04.022

[20] Oguntunde P, Fosu M, Ajayi A, Giesen N. Effect of charcoal production on maize yield, chemical properties and texture of soil. Biology and Fertility of Soils. 2004;**39**:295-299

[21] Rondon M, Lehmann J, Ramirez J, Hurtado M. Biological nitrogen fixation by common beans (*Phaseolus vulgaris* L.) increases with bio-char additions. Biology and Fertility of Soils. 2007;**43**:699-708

[22] Yamato M, Okimori Y, Wibowo IF, Anshiori S, Ogawa M. Effects of the application of charred bark of *Acacia mangium* on the yield of maize, cowpea and peanut, and soil chemical properties in South Sumatra, Indonesia. Soil Science and Plant Nutrition. 2006;**52**:489-458

[23] Genesio L, Miglietta F, Baronti S, Vaccari FP. Biochar increases vine-yard productivity without affecting grape quality: Results from a four years field experiment in Tuscany. Agriculture, Ecosystems and Environment. 2015;**201**:20-25

[24] Chan KY, van Zwieten L, Meszaros I, Downie A, Joseph S. Using poultry litter biochars as soil amendments. Australian Journal of Soil Research. 2008;**46**:437-444

[25] Steiner C, Teixeira WG, Lehmann J, Nehls T, Macedo JLV, Blum WEH, et al. Long-term effects of manure, charcoal and mineral fertilization on crop production and fertility on a highly weathered central Amazonian upland soil. Plant and Soil. 2007;**291**:275-290

[26] Spokas KA, Cantrell KB, Novak JM, Archer DA, Ippolito JA, Collins HP, et al. Biochar: A synthesis of its agronomic impact beyond carbon sequestration. Journal of Environmental Quality. 2012;**41**:973-989

[27] Petter FA, Madari BE, de Silva MAS, Cameiro MAC, Carvalho MTM, Ben HM Jr, et al. Soil fertility and upland rice yield after biochar application in the. Cerrado Pesq agropec bras Brasília. 2012;**47**(5):699-706

[28] Liu X, Zhang A, Ji C, Joseph S, Bian R, Li L, et al. Biochar's effect on crop productivity and the dependence on experimental conditions: A meta-analysis of literature data. Plant and Soil. 2013;**373**:583-594

[29] Hossain MK, Strezov V, Chan KY, Nelson PF. Agronomic properties of wastewater sludge biochar and bioavailability of metals in production of cherry tomato (*Lycopersicon esculentum*). Chemosphere. 2010;**78**:1167-1171

[30] Brantley KE et al. Pine woodchip biochar impacton soil nutrient concentrations and corn yield in a silt loam in the mid-southern U.S. Agriculture. 2015;**5**(1):30-47

[31] Ndor E, Ogara JI, Bako DA, Osuagbalande JA. Effect of biochar on macronutrients release and plant growth on degraded soil of Lafia, Nasarawa State, Nigeria. Asian Research Journal of Agriculture. 2016;**2**(3):1-8

[32] Muhammad FQ, Muhammed A, Suhan D, Muhammad KS, Muhammad AA. Effects of various biochars on seed germination and carbon mineralization in an alkaline soil. Pakistan Journal of Agricultural Sciences. 2015;**51**(4):977-982

[33] Fru BS, Francis NA, Angwafo TE, Precillia TN. Waterleaf (*Talinum triangulare*) response to biochar application in a humid-tropical forest soil. Journal of Soil Science and Environmental Management. 2017;**8**(5):95-103

[34] Kammann C, Ratering S, Eckchard C, Muller C. Biochar and hydrochar effects on greenhouse gas fluxes from soils. Journal of Environmental Quality. 2015;**41**:1052-1066

[35] Bamboo Biochar—Bio-organic Fertilizer. Biochar. 2019. Available from: https://biochar.co.nz/component/jce/?view=popup&tmpl=component&img=images/BIOCHAR-FILES/BBP-NO.2-Diagram.jpg&title=Process_Diagram&width=500 [Accessed: 22 April 2019]

[36] Subedi R, Kammann C, Pelissetti S, Taupe N, Bertora C, Monaco S, et al. Does soil amended with biochar and hydrochar reduce ammonia emissions following the application of pig slurry? European Journal of Soil Science. 2015;**66**:1044-1053

[37] Raghunath S, Chiara B, Laura Z, Carlo G. Crop response to soils amended with biochar: Expected benefits and unintended risks. Italian Journal of Agronomy. 2017;**12**:161-173

[38] Tariq MI, Afzal S, Hussain I, Sultana N. Pesticides exposure in Pakistan: A review. Environment International. 2007;**33**:1107-1122

[39] Atreya K. Health costs from short-term exposure to pesticides in Nepal. Social Science & Medicine. 2008;**67**:511-519

[40] Bale JS, van Lenteren JC, Bigler F. Biological control and sustainable food production. Philosophical Transactions of the Royal Society of London. Series B, Biological Sciences. 2008;**363**(1492):761-776

[41] Egamberdieva D, Wirth S, Behrendt U, Abd Allah EF, Berg G. Biochar treatment resulted in a combined effect on soybean growth promotion and a shift in plant growth promoting rhizobacteria. Frontiers in Microbiology. 2016;**7**:209. DOI: 10.3389/fmicb.2016.00209

[42] Berg G. Plant-microbe interactions promoting plant growth and health: Perspectives for controlled use of microorganisms in agriculture. Applied Microbiology and Biotechnology. 2009;**84**(1):11-18

[43] Abiala MA, Popoola OO, Olawuyi OJ, Oyelude JO, Akanmu AO, Killani AS, et al. A review: Harnessing the potentials of vesicular arbuscular mycorrhizal (VAM) fungi to plant growth. International Journal of Pure and Applied Sciences and Technology. 2013;**14**(2):61-79

[44] Olawuyi OJ, Odebode AC, Olakojo SA, Popoola OO, Akanmu AO, Izenegu JO. Host-pathogen interaction of maize (*Zea mays* L.) and *Aspergillus niger* as influenced by *Arbuscular mycorrhizal* fungi (*Glomus deserticola*). Archives of Agronomy and Soil Science. 2014;**60**(11):1577-1591

[45] Chukwuka KS, Okechukwu RU, Umukoro BO, Obiakara MC. *Arbuscular mycorrhiza* fungi, NPK (15,15,15) and cow dung interaction in sustainable cassava production and food security. Advances in Plants & Agricultural Research. 2017;**7**(4):00262

[46] Akanmu AO, Popoola AR, Adebare G, Abiala MA, Odebode AC, Adeoye GO. Effectiveness of some plant extracts in the management of tomato pith necrosis caused by *Pseudomonas*

corrugata. International Journal of Organic Agriculture Research and Development. 2012;**5**:127-140

[47] Akanmu AO, Abiala AM, Aiyelaagbe OO, Olaoluwa OO, Odebode AC. Inhibition of *Fusarium* pathogens in millet by extracts of *Jatropha curcas* and *Mangifera indica*. International Journal of Plant Biology & Research. 2014;**2**(1):1007

[48] Akanmu AO, Olawuyi OJ, Bello OB, Akinbode OA, Aroge T, Oyewole B, et al. Genotypic variations in the inhibitory potentials of four combined botanicals on mycelia growth of *Macrophomina phaseolina* of cowpea [*Vigna unguiculata* (L) Walp.]. Tropical Plant Research. 2015;**2**(3):257-263

[49] Wuddivira MN, Stone RJ, Ekwue EI. Structure stability of humid tropical soils as influenced by manure incorporation and incubation duration. Soil Science Society of America Journal. 2009;**73**:1353-1360

[50] Smith P, Gregory PJ, Van Vuuren D, Obersteiner M, Havlik P, Rounsevell M, et al. Competition for land. Philosophical transactions of the Royal Society of London. Series B, Biological sciences. 2010;**365**:2941-2957

[51] Atkinson CJ, Fitzgerald JD, Hipps NA. Potential mechanisms for achieving agricultural benefits from biochar application to temperate soils: A review. Plant and Soil. 2010;**337**:1-18

[52] Noguera D, Rondon M, Laossi KR, Hoyos V, Lavelle P, Cruz de Carvalho MH, et al. Contrasted effect of biochar and earthworms on rice growth and resource allocation in different soils. Soil Biology and Biochemistry. 2010;**42**:1017-1027

[53] Van Der Heijden MG, Bardgett RD, Van Straalen NM. The unseen majority: Soil microbes as drivers of plant diversity and productivity in terrestrial ecosystems. Ecological Letters. 2008;**11**:296-310

[54] Birk JJ, Steiner C, Teixeira WC, Zech W, Glaser B. Microbial response to charcoal amendments and fertilization of a highly weathered tropical soil. In: Woods WI, Teixeira WG, Lehmann J, Steiner C, Winkler Prins AMGA, Rebellato L, editors. Amazonian Dark Earths: Wim Sombroek's Vision. Berlin: Springer; 2009. pp. 309-324

[55] Schnee LS, Knauth S, Hapca S, Otten W, Eickhorst T. Analysis of physical pore space characteristics of two pyrolytic biochars and potential as microhabitat. Plant and Soil. 2016;**408**(1-2):357-368. DOI: 10.1007/s11104-016-2935-9

[56] Verheijen FGA, Jeffery S, Bastos AC, van der Velde M, Diafas I. Biochar Application to Soils - A Critical Scientific Review of Effects on Soil Properties, Processes and Functions. EUR 24099 EN. Luxembourg: Office for the Official Publications of the European Communities; 2009. pp. 149

[57] Shackley S, Ruysschaert G, Zwart K, Glaser B. Biochar in European Soils and Agriculture Science and Practice. Routledge; 2016. 302 pages|95 B/W Illus. pp. 90-94

[58] Rouquerol J, Avnir D, Fairbridge CW, Everett DH, Haynes JM, Pernicone N, et al. Recommendations for the characterization of porous solids (technical report). Pure and Applied Chemistry. 1994;**66**(8):1739-1758

[59] Gray M, Johnson MG, Dragila MI, Kleber M. Water uptake in biochars: The roles of porosity and hydrophobicity. Biomass and Bioenergy. 2014;**61**:196-205

[60] Novak JM, Busscher WJ, Watts DW, Amonette JE, Ippolito JA, Lima IM, et al. Biochars impact on soil-moisture storage in an ultisol and two aridisols. Soil Science. 2012;**177**:310-320

[61] Barnes RT, Gallagher ME, Masiello CA, Liu Z, Dugan B. Biochar-induced changes in soil hydraulic conductivity and dissolved nutrient fluxes constrained by laboratory experiments. PLoS One. 2014;**9**(9):e108340

[62] Hille M, Den Ouden J. Charcoal and activated carbon as adsorbate of phytotoxic compounds—A comparative study. Oikos. 2005;**108**:202-207

[63] Chen B, Zhou D, Zhu L. Transitional adsorption and partition of non-polar and polar aromatic contaminants by biochars of pine needles with different pyrolytic temperatures. Environmental Science & Technology. 2008;**42**:5137-5143

[64] Downie A, Crosky A, Munroe P. Physical properties of biochar. In: Lehmann J, Joseph S, editors. Biochar for Environmental Management: Science and Technology. London, UK: Earthscan; 2009. pp. 13-32

[65] Ahmad M, Rajapaksha AU, Lim JE, Zhang M, Bolan N, Mohan D, et al. Biochar as a sorbent for contaminant management in soil and water: A review. Chemosphere. 2014;**99**:19-33

[66] Thies J, Rilling M, Graber ER. Biochar effects on the abundance, activity and diversity of the soil biota. In: Lehmann J, Joseph S, editors. Biochar for Environmental Management: Science and Technology. 2nd ed. London, UK: Earthscan; 2015

[67] Warnock DD, Lehmann J, Kuyper TW, Rilling MC. Mycorrhizal responses to biochar in soil—Concepts and mechanisms. Plant and Soil. 2007;**300**:9-20

[68] Graber ER, Harel YM, Kolton M, Cytryn E, Silber A, David DR, et al. Biochar impact on development and productivity of pepper and tomato grown in fertigated soilless media. Plant and Soil. 2010;**337**:481-496

[69] Kolton M, Harel YM, Pasternak Z, Graber ER, Elad Y, Cytryn E. Impact of biochar application to soil on the root-associated bacterial community structure of fully developed greenhouse pepper plants. Applied and Environmental Microbiology. 2011;**14**:4924-4930

[70] Fischer D, Glaser B. Synergisms between compost and biochar for sustainable soil amelioration. In: Sunil K, Bharti A, editors. Management of Organic Waste. Rijeka, Croatia: In Tech; 2012. pp. 167-198. DOI: 10.5772/31200

[71] Schulz H, Dunst G, Glaser B. Positive effects of composted biochar on plant growth and soil fertility. Agronomy for Sustainable Development. 2013;**33**(4):817-827

[72] Burketova L, Trda L, Ott PG, Valentova O. Bio-based resistance inducers for sustainable plant protection against pathogens. Biotechnology Advances. 2015;**33**(6), (Part 2):994-1004

[73] Matsubara YI, Hasegawa N, Fukui H. Incidence of *Fusarium* root rot in *Asparagus* seedlings infected with *Arbuscular mycorrhizal* fungus as affected by several soil amendments. Journal of the Japanese Society for Horticultural Science. 2002;**71**:370-374

[74] De Tender CA, Debode J, Vandecasteele B, D'Hose T, Cremelie P. Biological, physicochemical and plant health responses in lettuce and strawberry in soil or peat amended with biochar. Applied Soil Ecology. 2016;**107**:1-12. DOI: 10.1016/j.apsoil.2016.05.001

[75] Yao Q, Liu J, Yu Z, Li Y, Jin J, Liu X, et al. Three years of biochar amendment alters soil physiochemical properties and fungal community composition in a black soil of Northeast China. Soil Biology and

Biochemistry. 2017;**110**:56-67. DOI: 10.1016/j.soilbio

[76] Akiyama K, Matsuzaki KI, Hayashi H. Plant sesquiterpenes induce hyphal branching in *Arbuscular mycorrhizal* fungi. Nature. 2005;**435**:824-827

[77] IPCC. Climate Change 2007: Synthesis Report. In: Core Writing Team, Pachauri RK, Reisinger A, editors. Contribution of Working Groups I, II and III to the Fourth Assessment Report of the Intergovernmental Panel on Climate Change. Geneva, Switzerland: IPCC; 2007. 104 pp

[78] Mukome FND, Six J, Parikh SJ. The effects of walnut shell and wood feedstock biochar amendments on greenhouse gas emissions from a fertile soil. Geoderma. 2013;**200-201**:90-98

[79] Hanson PJ, Edwards NT, Garten CT, Andrews JA. Separating root and soil microbial contributions to soil respiration: A review of methods and observations. Biogeochemistry. 2000;**48**(1):115-146. DOI: 10.1023/a:1006244819642

[80] Zimmerman AR, Gao B, Ahn MY. Positive and negative carbon mineralization priming effects among a variety of biochar-amended soils. Soil Biology and Biochemistry. 2011;**43**:1169-1179

[81] Joseph SD, Camps-Arbestain M, Lin Y, Munroe P, Chia CH, Hook J, et al. An investigation into the reactions of biochar in soil. Australian Journal of Soil Research. 2010;**48**:501-515

[82] Wang J, Chen Z, Xiong Z, Chen C, Xu X, Zhou Q, et al. Effects of biochar amendment on greenhouse gas emissions, net ecosystem carbon budget and properties of an acidic soil under intensive vegetable production. Soil Use and Management. 2015;**31**:375-383

[83] Gillam KM, Zebarth BJ, Burton DL. Nitrous oxide emissions from denitrification and the partitioning of gaseous losses as affected by nitrate and carbon addition and soil aeration. Canadian Journal of Soil Science. 2008;**88**:133-143

[84] Yanai Y, Toyota K, Okazaki M. Effect of charcoal addition on N_2O emissions from soil resulting from rewetting air-dried soil in short-term laboratory experiments. Soil Science & Plant Nutrition. 2007;**53**:181-188

[85] Stewart CE, Zheng J, Botte J, Cotrufo MF. Co-generated fast pyrolysis biochar mitigates green-house gas emissions and increases carbon sequestration in temperate soils. GCB Bioenergy. 2013;**5**:153-164

[86] Spokas K, Novak J, Venterea R. Biochar's role as an alternative N-fertilizer: Ammonia capture. Plant and Soil. 2012;**350**:35-42

[87] Jones DL, Rousk J, Edwards-Jones G, DeLuca TH, Murphy DV. Biochar-mediated changes in soil quality and plant growth in a three year field trial. Soil Biology and Biochemistry. 2012;**45**:113-124

[88] Karhu K, Mattila T, Bergstrom L, Regina K. Biochar addition to agricultural soil increased CH_4 uptake and water holding capacity—Results from a short-term pilot field study. Agriculture, Ecosystems and Environment. 2011;**140**:309-313

[89] Van Zwieten L, Kimber S, Morris S, Chan K, Downie A, Rust J, et al. Effects of biochar from slow pyrolysis of papermill waste on agronomic performance and soil fertility. Plant and Soil. 2010;**327**:235-246

[90] Bruun EW, Petersen CT, Hansen E, Holm JK, Hauggaard-Nielsen H. Biochar amendment to coarse sandy subsoil improves root growth and increases water retention. Soil Use and Management. 2014;**30**:109-118

[91] Quilliam RS, Rangecroft S, Emmett BA, Deluca TH, Jones DL. Is biochar a source or sink for polycyclic aromatic hydrocarbon (PAH) compounds in agricultural soils? GCB Bioenergy. 2012;**5**:96-103

[92] Rondon M, Ramirez JA, Lehmann J.. Charcoal additions reduce net emissions of greenhouse gases to the atmosphere. Volume 208. In: Proceedings of the 3rd USDA Symposium on Greenhouse Gases and Carbon Sequestration in Agriculture and Forestry; Maryland, USA; 2005

[93] Zhang A, Cui L, Pan G, Li L, Hussain Q, Zhang X, et al. Effect of biochar amendment on yield and methane and nitrous oxide emissions from a rice paddy from Tai Lake plain, China. Agriculture, Ecosystems and Environment. 2010;**139**:469-475

[94] Zhang A, Liu Y, Pan G, Hussain Q, Li L, Zheng J, et al. Effect of biochar amendment on maize yield and greenhouse emissions from a soil organic carbon poor calcareous loamy soil from Central China plain. Plant and Soil. 2012;**351**:263-275

[95] Cayuela ML, Jeffery S, van Zwieten L. The molar H: Corg ration of biochar is a key factor in mitigating N_2O emissions from soil. Agriculture, Ecosystems and Environment. 2015;**202**:135-138

[96] Fan C, Hao C, Bo L, Zhengqin X. Biochar reduces yield-scaled emissions of reactive nitrogen gases from vegetable soils across China. Biogeosciences. 2017;**14**:2851-2863

[97] Le Mer J, Roger P. Production, oxidation, emission and consumption of methane by soils: A review. European Journal of Soil Biology. 2001;**37**:25-50

[98] Topp E, Pattey E. Soils as sources and sinks for atmospheric methane. Canadian Journal of Soil Science. 1997;**77**:167-178

[99] Castro MS, Steudler PA, Melillo JM. Factors controlling atmospheric methane consumption by temperate forest soils. Global Biogeochemical Cycles. 1995;**9**:1-10

[100] Wang J, Pan X, Liu Y, Zhang X, Xiong Z. Effects of biochar amendment in two soils on greenhouse gas emissions and crop production. Plant and Soil. 2012;**360**:287-298

[101] Knoblauch C, Maarifat AA, Pfeiffer EM, Haefele SM. Degradability of black carbon and its impact on trace gas fluxes and carbon turnover in paddy soils. Soil Biochem. 2011;**43**:1768-1778

[102] Castaldi A. Impact of biochar application to a Mediterranean wheat crop on soil microbial activity and greenhouse gas fluxes. Chemosphere. 2011;**85**:1464-1471

[103] Scheer C, Grace P, Rowlings D, Kimber S, Van Zwieten L. Effect of biochar amendment on the soil-atmosphere exchange of greenhouse gases from an intensive subtropical pasture in northern New South Wales, Australia. Plant and Soil. 2011;**345**:47-58

[104] Liu Y, Yang M, Wu Y, Wang H, Chen Y, Wu W. Reducing CH_4 and CO_2 emissions from waterlogged paddy soil with biochar. Journal of Soils and Sediments. 2011;**11**:930-939

[105] Ameloot N, De Neve S, Jegajeevagan K, Yildiz G, Buchan D, Funkuin YN. Short-term CO_2 and N_2O emissions and microbial properties of biochar amended sandy loam soils. Soil Biology and Biochemistry. 2013;**57**:401-410

[106] Van Zwieten L, Kimber S, Morris S, Chan KY, Downie A, Rust J, et al. Effects of biochar from slow pyrolysis of papermill waste on agronomic performance and soilfertility. Plant and Soil. 2010;**327**:235-246

[107] Laine MM, Jùrgensen KS. Effective and safe composting of chlorophenol-contaminated soil in pilot scale. Environmental Science & Technology. 1997;**31**:371-378

[108] Joyce JF, Sato C, Cardenas R, Surampalli RY. Composting of polycyclic aromatic hydrocarbons in simulated municipal solid waste. Water Environment Research. 1998;**70**:356-361

[109] Wei S, Zhu J, Zhou QX, Zhan J. Fertilizer amendment for improving the phytoextraction of cadmium by a hyperaccumulator *Rorippa globosa* (Turcz.) Thell. Journal of Soils and Sediments. 2011;**11**:915-922

[110] Wang T, Camps Arbestain M, Hedley M, Bishop P. Chemical and bioassay characterisation of nitrogen availability in biochar produced from diary manure and biosolids. Organic Geochemistry. 2012;**51**:45-54

[111] Asemoloye MD, Jonathan SG, Jayeola AA, Ahmad R. Mediational influence of spent mushroom compost on phytoremediation of black-oil hydrocarbon polluted soil and response of *Megathyrsus maximus* Jacq. Journal of Environmental Management. 2017;**200**:253-262

[112] Asemoloye MD, Ahmad R, Jonathan SG. Synergistic action of rhizospheric fungi with *Megathyrsus maximus* root speeds up hydrocarbon degradation kinetics in oil polluted soil. Chemosphere. 2017;**184**:1-12

[113] Ji L, Wan Y, Zheng S, Zhu D. Adsorption of tetracycline and sulfamethoxazole on crop residue-derived ashes: Implication for the relative importance of black carbon to soil sorption. Environmental Science & Technology. 2011;**45**:5580-5586

[114] Uchimiya M, Lima IM, Klasson KT, Wartelle LH. Contaminant immobilization and nutrient release by biochar soil amendment: Roles of natural organic matter. Chemosphere. 2010;**80**:935-940

[115] Singh B, Singh BP, Cowie A. Characterization and evaluation of biochars for their application as a soil amendment. Australian Journal of Soil Research. 2010;**48**:516-525

[116] Mukherjee A, Lal R, Zimmerman AR. Effects of biochar and other amendments on the physical properties and greenhouse gas emissions of an artificially degraded soil. Science of The Total Environment. 2014;**487**:26-36

[117] Barrow CJ. Biochar: Potential for countering land degradation and for improving agriculture. Applied Geography. 2012;**34**:21-28

[118] Cayuela ML, Van Zwieten L, Singh BP, Jeffery S, Roig A, Sánchez-Monedero MA. Biochar's role in mitigating soil nitrous oxide emissions: A review and meta-analysis. Agriculture, Ecosystems and Environment. 2014;**191**:5-16

[119] Lin Y, Munroe P, Joseph S, Henderson R, Ziolkowski A. Water extractable organic carbon in untreated and chemical treated bio-chars. Chemosphere. 2012;**87**:151-157. DOI: 10.1016/j.chemosphere.2011.12.007

[120] Ameloot N, Graber ER, Verheijen FGA, De Neve S. Interactions between biochar stability and organisms: Review and research needs. European Journal of Soil Science. 2013;**64**:379-390

[121] Ameloot N, Maenhout P, De Neve S, Sleutel S. Biochar induced N_2O emission reductions after field incorporation in a loam soil. Geoderma. 2016;**267**:10-16

[122] Li J, Dai J, Liu G, Zhang H, Gao Z, Fu J, et al. Biochar from microwave

pyrolysis of biomass: A review. Biomass and Bioenergy. 2016;**94**:228-244

[123] Windeatt JH, Ross AB, Williams PT, Forster PM, Nahil MA, Singh S. Characteristics of biochars from crop residues: Potential for carbon sequestration and soil amendment. Journal of Environmental Management. 2014;**15**:189-197

[124] Novak JM, Ro K, Ok YS, Sigua G, Spokas K, Uchimiya S, et al. Biochars multifunctional role as a novel technology in the agricultural, environmental, and industrial sectors. Chemosphere. 2016;**142**:1-3

[125] Karami N, Clemente R, Moreno-Jiménez E, Lepp N, Beesley L. Efficiency of green waste compost and biochar soil amendments for reducing lead and copper mobility and uptake to ryegrass (*Lolium perenne*). Journal of Hazardous Materials. 2011;**191**:41-48

[126] Wenzel W. Rhizosphere processes and management in plant-assisted bioremediation (phytoremediation) of soils. Plant and Soil. 2009;**321**:385-408. DOI: 10.1007/s11104-008-9686-1

[127] Clemente R, Walker DJ, Pardo T, Martínez-Fernández D, Bernal MP. The use of a halophytic plant species and organic amendments for the remediation of a trace elements-contaminated soil under semi-arid conditions. Journal of Hazardous Materials. 2012;**223-224**:63-71. DOI: 10.1016/j.jhazmat

[128] Beesley L, Moreno-Jiménez E, Gomez-Eyles JL, Harris E, Robinson B, Sizmur T. A review of biochars' potential role in the remediation, revegetation and restoration of contaminated soils. Environmental Pollution. 2011;**159**:3269-3282. DOI: 10.1016/j.envpol

[129] Prendergaast-Miller MT, Duvall M, Sohi SP. Localisation of nitrate in the rhizosphere of biochar-amended soils. Soil Biology and Biochemistry. 2011;**43**:2243-2246

[130] Prendergast-Miller MT, Duvall M, Sohi SP. Biochar-root interactions are mediated by biochar nutrient content and impacts on soil nutrient availability. European Journal of Soil Science. 2014;**65**:173-185. DOI: 10.1111/ejss.12079

[131] Camps-Arbestain M, Saggar S, Leifeld J. Environmental benefits and risks of biochar application to soil. Agriculture, Ecosystems and Environment. 2014;**191**:1-4

[132] Nelissen V, Saha BK, Ruysschaert G, Boeckx P. Effect of different biochar and fertilizer types on N_2O and NO emissions. Soil Biology and Biochemistry. 2014;**70**:244-255. DOI: 10.1016/j.soilbio.2013.12.026

[133] Gul S, Whalen JK, Thomas BW, Sachdeva V. Physico-chemical properties and microbial responses in biochar-amended soils: Mechanisms and future directions. Agriculture, Ecosystems and Environment. 2015;**206**:46-49

[134] Harel YM, Elad Y, Rav-David D, Borenstein M, Shulchani R, Lew B, et al. Biochar mediates systemic response of strawberry to foliar fungal pathogens. Plant and Soil. 2012;**357**:245-257

[135] George C, Kohler J, Rillig MC. Biochars reduce infection rates of the root-lesion nematode *Pratylenchus penetrans* and associated biomass loss in carrot. Soil Biology and Biochemistry. 2016;**95**:11-18

[136] DeLuca TH, Gundale MJ, Mac Kenzie MD, Jones DL. Biochar effects on soil nutrient transformations (chapter 15). In: Lehmann J, Joseph S, editors. Biochar for Environmental Management: Science, Technology and

Implementation. 2nd ed. Abingdon, UK: Routledge; 2015. pp. 421-455

[137] Subedi R, Taupe N, Ikoyi I, Bertora C, Zavattaro L, Schmalenberger A, et al. Chemically and biologically-mediated fertilizing value of manure-derived biochar. Science of The Total Environment. 2016;**550**:924-933

[138] Subedi R, Taupe N, Pelissetti S, Petruzzelli L, Bertora C, Leahy JJ, et al. Greenhouse gas emissions and soil properties following amendment with manure-derived biochars: Influence of pyrolysis temperature and feedstock type. Journal of Environmental Management. 2016;**166**:73-83

[139] Drake JA, Cavagnaro TR, Cunningham SC, Jackson WR, Patti AF. Does biochar improve establishment of tree seedlings insaline sodic soils? Land Degradation and Development. 2016;**27**:52-59

[140] Cheng CH, Lehmann J. Ageing of black carbon along a temperature gradient. Chemosphere. 2009;**75**:1021-1027

[141] Hartley W, Dickinson NM, Riby P, Lepp NW. Arsenic mobility in brown field soils amended with green waste compost or biochar and planted with *Miscanthus*. Environmental Pollution. 2009;**157**:2654-2662

[142] Fellet G, Marmiroli M, Marchiol L. Elements uptake by metal accumulator species grown on mine tailings amended with three types of biochar. Science of the Total Environment Volumes. 2014;**468-469**:598-608

[143] Houben D, Evrard L, Sonnet P. Beneficial effects of biochar application to contaminated soils on the bioavailability of Cd, Pb and Zn and the biomass production of rapeseed (*Brassica napus* L.). Biomass and Bioenergy. 2013;**57**:196-204

[144] Ippolito JA, Laird DA, Busscher WJ. Environmental benefits of biochar. Journal of Environl Quality. 2012;**41**:967-972

[145] Ippolito JA, Spokas KA, Novak JM, Lentz RD, Cantrell KB. Biochar elemental composition and factors influencing nutrient retention. In: Lehmann J, Joseph S, editors. Biochar for Environmental Management: Science, Technology and Implementation. 2nd ed. Abingdon, UK: Routledge; 2015. pp. 139-163

[146] Kuppusamy S, Thavamani P, Megharaj M, Venkateswarlu K, Naidu R. Agronomic and remedial benefits and risks of applying biochar to soil: Current knowledge and future research directions. Environment International. 2016;**87**:1-12

[147] Liang B, Lehmann J, Sohi SP, Thies JE, O'Neill B, Trujillo L, et al. Black carbon affects the cycling of non-black carbon in soil. Organic Geochemistry. 2010;**41**:206-213

[148] Chintala R, Schumacher TE, Kumar S, Malo DD, Rice J, Bleakley B, et al. Molecular characterization of biochar materials and their influence on microbiological properties of soil. Journal of Hazardous Materials. 2014;**279**:244-256

[149] Thies EJ, Rilling MC. Characteristics of biochar: Biological properties. In: Lehmann J, Joseph S, editors. Biochar for Environmental Management: Science and Technology. London: Earthscan; 2009. pp. 85-102

[150] Steinbeiss S, Gleixner G, Antonietti M. Effect of biochar amendment on soil carbon balance and soil microbial activity. Soil Biology and Biochemistry. 2009;**41**:1301-1310. DOI: 10.1016/j.soilbio.2009.03.016

[151] Anderson CR, Condron LM, Clough TJ, Fiers M, Stewart A, Hill RA. Biochar induced soil microbial

community change: Implications for biogeochemical cycling of carbon, nitrogen and phosphorus. Pedobiologia. 2011;**54**:309-320

[152] Jha P, Biswas AK, Lakaria BL, Rao AS. Biochar in agriculture—Prospects and related implications. Current Science. 2010;**99**:1218-1225

[153] Cross A, Sohi SP. The priming potential of biochar products in relation to labile carbon contents and soil organic matter status. Soil Biology and Biochemistry. 2011;**43**:2127-2134

[154] Swaine M, Obrike R, Clark JM, Shaw LJ. Biochar alteration of the sorption of substrates and products in soil enzyme assays. Applied and Environmental Soil Science. Article ID 968682. 2013. DOI: 10.1155/2013/968682

[155] Asemoloye MD, Ahmad R, Jonathan SG. Synergistic rhizosphere degradation of γ-hexachlorocyclohexane (Lindane) through the combinatorial plant-fungal action. PLoS One. 2017;**12**(7):e-0183373

[156] Schmalenberger A, Fox A. Bacterial mobilization of nutrients from biochar amended soils. Advances in Applied Microbiology. 2016;**94**:109-109

Chapter 5

Role of Soil Microbes on Crop Yield against Edaphic Factors of Soil

Shishir Raut

Abstract

Soil degradation is one or the other form and consequent decline in soil productivity which has been the experience of the farmer since ancient times. As population pressure on agricultural land increases, concerns for ensuring sustained agricultural productivity of soils are being voiced more vociferously now. Chemical degradation of soils takes place as these accumulate soluble salts or toxic elements in amounts deleterious for plant growth or their chemical properties are so transformed as to adversely affect their productivity. The losses in soil productivity may also be accompanied by ecological obliteration and environmental degradation of the whole area. Continuous use of inorganic fertilizers coupled with depletion of organic matter results in deterioration of soil structure and soil productivity. It leads to reduce input/output ratio unless soils are replenished with organic matter through green manure, farm yard manure (FYM), compost or through microbial activity. Due to repeated application of microbes like blue green algae (BGA), biofertilizer soil organic carbon content is not only maintained but enriched too. The increase in carbon content of saline soil of Andhra Pradesh (India) has been shown to be up to 22%. The microbial polysaccharides are regarded as the most important natural products in the formation and stabilization of soil aggregates.

Keywords: soil salinity, alkalinity, acidity, organic matter, microbes

1. Introduction

Increasing pace of production to meet the demands of growing population is resulting in overexploitation of both renewable and non-renewable resources and accumulation of contaminants in environment, as wastes. Amongst other things, more emphasis is now being placed on recycling technology to prevent the depletion of resources and to limit environmental degradation due to overloading of soil and atmosphere with residues/byproducts of chemical and physical processing industries. Initially, this technology dealt with protection of environment through waste water treatment, disposal of human excreta, use of crop residue for improving soil fertility, etc. The field has now widened to control environmental pollution, waste land reclamation and conversion of wastes into industrially valuable products. In this context, bioconversions/bio transformations through microbes are attracting greater attention, since enzymes carry out very specific reactions under mild conditions, larger water insoluble molecules can be transformed and biogeochemical cycling do not require external energy inputs [1].

Soil and water conservation has been practiced extensively by settled agriculturist traditionally to maintain the productivity of any country. When the population pressures increase, the farming systems break down as there is an upper limit for any land water system to support increasing number of human beings and livestock without affecting the ecological balance. Over the years the population pressure has been increased and per capita availability of arable land is decreased. Thus soil and land resources, in recent years, are under tremendous pressure with highly conflicting and competing demands of rising population. The increased claims on land for settlement, urban growth, industrialization and other development purposes apart from increased demands for food, fodder and fibre, has set in an imprudent trend of land utilization with disregard to risks of damage to the ecosystem. The apathy for the harmonious use of land in accordance with potentialities and capabilities of soils has given rise to a multitude of serious problems. The major soil problems in some tropical and subtropical countries are salinity, alkalinity, waterlogging, acidity and soil fertility. Soil degradation thus refers to appreciable loss of productivity and is defined as a process which lowers the current and/or potential capability of soil to produce goods or services [2].

2. Chemical degradation of soils

Chemical degradation of soil is mediated through processes that induce leaching of bases, development of acidity, deficiencies of certain nutrient elements or accumulation of metallic ions in quantities toxic for plant growth. Reducing conditions of soil following prolonged waterlogging may generate toxic levels of organic constituents and certain metallic ions. It also includes the processes that help accumulate harmful levels of salts of various kinds in the soil.

2.1 Soil degradation through base unsaturation

Excessive removal of bases from surface soil leads to disorder in crop plants grown in highly acid soils met under different pedogenic environments. In high altitude soils of cold climate, soil acidity and base unsaturation is associated with soil forming processes normal to the given climate and local conditions. Apart from lime needs soil degradation is not a major threat to agricultural productivity in the above regions. In high rainfall tropical areas excessive soil acidity may assume discernible proportions in soil types like laterite, lateritic, red and even alluvial soils borne on different parent materials. In the high rainfall forest soils of the hilly region, high organic matter and rainfall tend to produce acidity and base unsaturation in soils with concomitant increase in exchangeable Al of the soil. Often local conditions like rise in water table along with high organic matter and rice cultivation may lead to high active iron in the soil with its content as high as 3%. Besides toxicity of iron, the crops may experience deficiency of phosphorus and zinc. Often toxic concentration of M_n may also be encountered in such soils [3]. All these factors, lead to loss in soil productivity.

2.2 Sources of salts in soil (Indian context)

2.2.1 Relic salts

Prior to the orogeny of the Himalayan ranges, the Tethys sea extended in a large part of the present Indo Gangetic plains, Western Rajasthan and the Kuchh. With uplift of the Himalayas the foredeep was filled up with detritus. Some of the sea salts precipitated during the process of deposition while other entrapped in the alluvium are present mainly in the arid parts of the country [3]. In the lower Himalayas the rock salt deposits of the pre-cambrian saline series in Himachal Pradesh and the salt range of Pakistan and numerous brine springs in the Shivalik region rich with as much as 3243 mg l^{-1} chlorides, are important sources of salts [3].

2.2.2 In-situ weathering

The only source of in-situ release of salts is weathering of minerals constituting the soil regolith below it. Its ample evidences are available in peninsular regions. The alluvial zone in the north is composed of strongly saline sedimentary rocks which originated during the Territory and Pleistocene times. Their decomposition is incomplete in the arid climate. Hence there are constant sources of salts under environments which favour their further break down. For example almost 44% of light minerals in sand dunes of Rajasthan contain orthoclase feldspars which undergo weathering and release minerals like illite and montmorillonite [4]. Salts are natural by products of such weathering.

2.2.3 Overland flows

Run off waters pick up salts on their way to natural depressions. In some areas, the natural settings are such that they have centripetal drainage which attracts salt bearing over land flows from the surrounding areas. Minor channels feeding these basins owe their salinity to salt releasing marine lithological formations [5]. Even rivers flowing in the regions pick up lot of salts. In Luni basin of Rajasthan, India flash flood deposited thick layer of sediments composed of very fine sandy to coarse silty material charged with salts over highly productive agricultural lands rendering them so saline that cultivation had to be abandoned [6].

2.2.4 Sub-terranean flux of salts

In the peninsular regions, much of salinity in valley lands is traced to salts travelling from uplands laterally along the interface between the soil and the underlying 'murrum' or through the porous 'murrum' itself. Irrigation canals and channels are on the ridge causing seepage water to pick-up salts and move down to valley lands. The morphogenesis of salt lakes and palayas in Western Rajasthan of India is traced to the confluence of prior drainage channels. Rain water sinks through their beds and flows subterraneously along buried channel patterns carrying soluble salts washed down from the catchment. Evaporation of water from the bed concentrates salts in the path raising their salinity to as much as 3.2 gkg^{-1} of salts [7].

2.2.5 Rain and wind borne salts

The rain in salt water originate either from strong winds that sweep over oceans or salts picked from dust storms causing salt content of rain water to vary with locality and season. Salt additions from rains and winds may thus constitute an important source of salts in soils [7, 8].

2.2.6 Tidal floods and sea water intrusion

Coastal regions experience tidal floods especially through backwater creeks and rivers which spill over lands during high tide. Sub-surface intrusion of sea water in coastal areas is threatening agriculture in those areas [8].

2.2.7 Irrigation waters

All irrigation waters, irrespective of their source contain some soluble salts (**Table 1**). Even rivers of the Indo-Gangetic system which are snow fed may carry salty sediments during rainy season and significant salt loads during the lean discharge period. But in the absence of any exit, even small additions of salts over long periods can render the soil saline. In many arid and semi-arid regions, where canal water is scarce or not available, ground water is the sole source of irrigation. Even tank and lakes of many areas carry large salt loads. Prolonged use of such water in low rainfall areas, where natural leaching of salts fall short of their input into the soil, may render irrigated soils saline.

2.2.8 Coastal saline soils

A major portion of coastal saline soils occurs in the deltaic regions of major rivers, for India, falling either into the Bay of Bengal or the Arabian sea. A relatively smaller area of coastal saline soils occurs as narrow strips of lands along the sea coast and along the water lakes such as Chilika lake in Odisha. The soils of deltaic

Irrigation source	**Surface waters**		
	EC (dS/m)	**pH**	**SAR**
Rivers			
Ganges system	142–647	7.6–8.4	—
Indus system	370–420	7.2–7.7	—
Krishna-Godavari	725–1392	—	—
Vedavathi, Karnataka	1900	8.8	12.8
Tanks			
Gosikere, Karnataka	1400	8.6	10.8
Etah (U.P.)	3752	9.0	—
Kanpur (U.P.)	1766	8.5	—
Canals			
Dodherde (Karnataka)	1400	8.6	10.8
Nannewa (Karnataka)	1900	8.6	26.8

Ground waters					
State	**No. of samples tested**	**% distribution in EC_{iw} (dS/m) classes**			
		Up to 3	**3–5**	**5–10**	**>10**
Punjab	12,500	68.0	20.0	8.0	4.0
Haryana	3637	58.5	17.5	13.2	10.8
U.P. (Aligarh)	390	78.4	18.7	2.9	—
A.P. (Coastal)	1082	78.2	16.1	5.7	—
Karnataka (Bijapur)	404	87.6	8.9	3.2	0.2
Gujarat (Ahmedabad-Kheda)	505	84.0	100	5.7	0.2

EC, electrical conductivity; SAR, sodium adsorption ratio; iw, irrigation water.

Table 1.
Salt load of some irrigation waters in India [8].

regions, usually have flat topography and finer texture, than the other types of coastal soils, depending on the geomorphology of flood plains; coarse texture soils may also be found in the deltaic regions. The coastal soils of deltaic region are usually formed from the indirect deposits of alluvial materials going to the sea and transported back by the tides and redeposited in the estuarian/deltaic regions. The coastal saline soils have saline ground water at shallow depth. Both the ground water and the soils are rich in chlorides and sulphates of sodium, magnesium and calcium. The soil salinity and the depth to ground water vary with the season. Soil salinities are maximum in dry season and minimum in monsoon months (Indian Society of Coastal Agricultural Research 1987;5:1-14). The clay minerals vary with the region. The pH of the soils usually varies from slightly acidic to slightly alkaline except that the soils with high content of pyritic materials become strongly acidic on drying. Some such acidic soils are present in the coastal regions of Kerala, Sundarbans delta of India and in Andaman and Nicobar islands. Many a time, the exchangeable sodium percentage (ESP) of coastal saline soils is more than 15, but because of high salt content, it does not show strong alkali soil characters. The soils under cultivation are deficient in available nitrogen and organic carbon varying from 0.1% to 1.0% (Indian Society of Coastal Agricultural Research 1990;8:61-78). The organic matter content and its humic components also differ in different landforms as shown in **Table 2**. The humic acid (H.A.) and fulvic acid (F.A.) fractions of organic matter for a coastal soil of West Bengal, are given in **Table 2**. Humus is the major soil organic matter component making up 75–805 of the total [10]. Fractionation of organic matter showed that the fraction of H.A. was the highest (0.31%) in depressed low (DL) soil and the fraction of F.A. was the lowest (0.10%) in the surface layer of the same soil. On the other hand, the F.A. fraction was the highest in non-cultivated deltaic (NCD) soil (0.12%) for which these soils were more capable of infiltration. DL soil with greater fraction of insoluble humic acid exhibited less cumulative infiltration. Mud flat (MUD) soils showed intermediate values (0.11%). In the lower soil layers also H.A. percentage was higher in the DL soils (0.27–0.29). The H.A./F.A. ratio decreased with depth (0.7–0.5 for NCD and 3.1–3.0 for DL land soils) [11, 12]. Presence of humic acid in soil generally

Name of soil	Total organic matter	H.A	F.A.	H.A/F.A. ratio
0–20 cm				
NCD	2.1	0.08	0.12	0.7
MUD	1.8	0.18	0.11	1.7
DL	1.1	0.31	0.10	3.1
20–40 cm				
NCD	1.9	0.08	0.13	0.6
MUD	1.1	0.16	0.10	1.6
DL	0.93	0.29	0.09	3.0
40–60 cm				
NCD	0.88	0.07	0.11	0.5
MUD	1.1	0.14	0.09	1.5
DL	0.87	0.27	0.09	3.0

NCD, non-cultivated deltaic; MUD, mudflat/mangrove; DL, depressed low land.

Table 2.
Humic acid, fulvic acid content of organic matter and their ratio [9].

decreases volumetric water content of soil. Decline in water repellency of soil is due to the presence of water soluble fulvic acid [13, 14]. In general, with increase in EC values, there was a decrease in organic carbon content. This may be attributed to the decrease in activity of organic matter sequestering organisms. The organic C percentage was high at EC values 4–4.5 (dS/m) which may be because of addition of F.Y.M. (**Figure 1**).

2.3 Alkalinity (sodicity) of soil and their reclamation

The distinguishing characteristics of a sodic soil are high exchangeable sodium percentage (ESP) sufficient to interfere with plant growth; high sodium adsorption ratio (SAR) of saturation extract; presence of large concentration of sodium carbonate type salts, and low permeability. In sodic soil the ESP is more than 15, EC_e (electrical conductivity of saturation extract) is less than 4 dS/m and pH of the saturation paste is usually more than 8.5. The work at the Central Soil Salinity

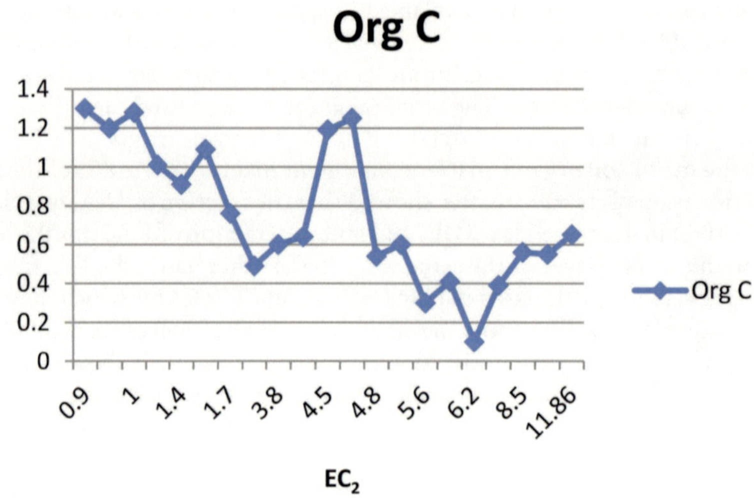

Figure 1.
Soil salinity and organic carbon relationship [9]. X, electrical conductivity values of 1:2 soil:water solution; Y: organic carbon %.

Soil		Saturation extract ($mmolL^{-1}$)	
pH_s	**10.3**	**Na^+**	**37.0**
EC_e	3.4	K^+	0.1
CEC [cmol (p+) kg^{-1}]	7.7	Ca^{2+}	1.4
Exchangeable [cmol (p+)kg^{-1}]		Mg^{2+}	0.6
Na^+	5.1	CO_3^{2-}	12.0
K^+	0.3	HCO_3^-	5.4
Ca^{2+}	1.2	Cl^{-1}	10.2
Mg^{2+}	0.7	SO_4^{2-}	12.6
ESP	66.2	SAR	37.0

Source: Report No. 8, Central Soil Salinity Research Institute, 1978, p. 72. pH_s refers to pH of saturation soil paste.

Table 3.
Characteristics of a typical sodic soil from Karnal, Haryana, India.

Gypsum** (t/ha)	Wheat cultivation at 1970–1971	Rice 1971	Wheat 1971–1972	Rice 1972	Wheat 1972–1973
0	0	4390	1520	7180	1350
7.5	1890	6210	2790	7230	1960
15.0	3490	6390	3450	7130	2430
22.5	4160	7080	3720	7170	2260
30.0	3790	6660	3980	7030	2740
CD (p = 0.05)	640	810	70	NS	500

Source: Soil Science 1970;127:79.
***Gypsum was applied to wheat in 1970.*

Table 4.
Effect of gypsum treatments on the yield of wheat and rice (kgha^{-1}).

Research Institute, Karnal, India suggests that for diagnostic purposes, pH 8.2 of the saturation paste may be taken as the lower limit of pH. In literature such soils have often been referred to as alkali soils. Characteristics of a typical sodic soil are presented in **Table 3**.

The problem of sodic soils is the high exchange sodium percentage. Obviously, the basic principle underlying reclamation of these soils is to adopt those ameliorative measures by which the exchangeable sodium will be replaced by calcium and the exchangeable sodium thus released as sodium salts will be leached out of the root zone. Because of low cost and easy availability, gypsum ($CaSO_4.2H_2O$) has been used widely and intensively as an amendment for reclamation. Hydraulic conductivity of sodic soils is very significantly increased by gypsum application and this result in increased yield of crops. An example of increase in crop yield by gypsum application is given in **Table 4**.

3. Soil degradation through water logging

Waterlogging of soils occurs when water balance of an area gets disturbed due to external inputs of water. Important source are heavy rains, overland flows, seepage from canals, tidal flooding especially through back water canals and coastal lakes. The soil situations favouring excess water concentration in a given area are basin type of topography with no natural outlet of water. Waterlogging creates anaerobic condition in soil which hampers the activity of beneficial soil microbes [14].

4. Environmental factors influencing microbial activity

Soil microflora, just like higher plants, depends entirely on soil for their nutrition and growth.

4.1 Water

Water is a major component of protoplasm in a microbial cell and is essential for growth. In the presence of excess water, say waterlogging, the environment becomes anaerobic because of lack of soil aeration, the aerobes becomes suppressed and inactive and anaerobic bacteria dominates. Nitrogen-fixing algae like *Tolypothrix tenuis*, *Scytonema cincinnatum* and *Hapalosiphon fontinalis* can be used in waterlogged areas which will fix nitrogen and will be useful to improve yield of crops (rice).

4.2 Temperature

Temperature is the most important factor influencing the biological processes and the microbial activity. The optimum temperature range at which a particular microorganism grows is narrow. Most of the soil organisms are mesophiles and grow well between 15 and 45°C.

4.3 Aeration

Microbes consume oxygen from soil air and give out carbon di oxide. Waterlogging reduces soil aeration and carbon di oxide is accumulated in soil which is toxic to the microbes.

4.4 Soil reaction

Bacteria, in general prefer near neutral to slightly alkaline reaction between pH 6.5 and 8.0; fungi grow in acidic reaction between pH 4.5 and 6.5.

4.5 Soil factor

A soil in bad physical condition as in degraded soil has not having good aeration and water supplying capacity which affect optimal microbial activity.

4.6 Microbial inoculation to ameliorate soil

Continuous use of inorganic fertilizers coupled with depletion of organic matter results in deterioration of soil structure and soil productivity. It leads to reduced input/output ratio unless soils are replenished with organic matter through green manure, FYM, compost or through microbial activity. Due to repeated application of BGA biofertilizer, soil organic carbon content is not only maintained but enriched too [15, 16]. The increase in carbon content of saline soils has been shown to be up to 22%. The microbial polysaccharides are regarded as the most important natural products in the formation and stabilization of soil aggregates. Presence of excess neutral soluble salts or high level of sodium in soils leads to rise in soil pH and soil finally become saline or sodic. These soils usually give poor crop yields or crops altogether fail. The crop failure is brought about either by nonavailability of plant nutrients or by the toxic effect of sodium ions per se. Repeated application of suitable BGA strains in such soils helps to bring down the level of soluble salts, pH towards neutrality and sodium content in exchange complex [17]. The cumulative effect of reduction in soil pH, electrical conductivity and exchangeable sodium, improvement of soil aggregation and permeability of air and water, together with enrichment of soil carbon content brings an overall improvement in soil health and thus productivity.

4.7 Crop response to algalization (in India)

Large numbers of field trials were conducted in different agroclimatic regions of India to assess the effect of algalization on rice yield. Agencies involved were state department of agriculture, All India Coordinated Rice Improvement Project (AICRIP), Hyderabad and progressive farmers [18]. In areas where chemical

State/organization	No. of trials	Yield (t/ha)		% increase over control
		Control	BGA	
Andhra Pradesh	1	3.64	4.44	21.9
AICRIP	20	3.04	3.37	10.8
Bihar	1	2.13	3.06	32.8
J & K		3.75	3.90	4.0
Madhya Pradesh	1	2.48	2.85	14.9
Maharashtra	161	3.05	3.91	28.1
Orissa	91	2.97	3.71	24.6
Punjab	1	5.04	5.27	4.5
Uttar Pradesh	1	3.29	3.82	16.1
Total trials	17			
Average	294	3.28	3.81	16.1

Table 5.
Yield of rice due to algalization in absence of inorganic nitrogen fertilizer [18].

State/organization	No. of trials	Nitrogen kg/ha	Average yield t/ha	% increase
J & K, M.P., U.P.	7	20 20 + BGA	3.78 4.15	9.74
AICRIP, Maharashtra	43	25 25 + BGA	3.71 4.09	10.21
J & K, Kerala, U.P.	15	30 30 + BGA	3.49 3.92	12.28
J & K, M.P., U.P.	7	40 40 + BGA	4.27 4.51	5.54
A.P., J & K, Maharashtra	29	50 50 + BGA	4.38 4.90	11.93
Kerala, M.P., U.P.	22	60 60 + BGA	3.87 4.34	11.93
Maharashtra	27	75 75 + BGA	4.46 4.88	9.50
Punjab, U.P.	6	80 80 + BGA	5.56 5.58	0.23
Kerala, U.P.	14	90 90 + BGA	3.64 3.99	9.77
A.P., Maharashtra, T.N.	46	100 100 + BGA	5.02 5.38	7.22
Kerala, Punjab, U.P.	15	120 120 + BGA	4.73 5.03	6.25
A.P.	1	150 150 + BGA	5.84 6.52	11.60
Total trials Average % increase	232			

Table 6.
Effect of algalization on the yield of rice at different levels of nitrogenous fertilizers [18].

N-fertilizers are not used for various reasons, algal inoculation enhances the crop yield with a minimum of 45 and a maximum of 32.8% in different places with an Indian average of 16.1% (**Table 5**). Even at the recommended levels of chemical nitrogen fertilizer being used in different areas application of BGA bio-fertilizer results in an increased yield of about 8.85%. Depending upon the level of nitrogen fertilizer and agro-climatic zones the yield increase varies from less than 1% to 12.28% (**Table 6**).

5. Conclusions

Use of microbes in resource management has been an age-old practice although the scientific reasoning for such practices has come to be known only recently. For instance, a common wetland plant, *Phragmites karka*, is traditionally used in rural Bengal (India) to make waste water suitable for fish culture. It has now been found that the detoxification is actually due to biodegradation of harmful elements by microorganisms associated with the roots of this plant. This knowledge led to worldwide use of such plants to recycle sewage, paper mill and distillery effluents, etc. for making water not only suitable for irrigation and aquaculture but also potable. The technology is known as 'root zone method'. The method is being used to treat effluents containing sulphur compounds, reactive dyes, ammonia, phosphates, chlorinated hydrocarbons and heavy metals. Similarly growing legumes with nonlegumes in the cropping system improves soil fertility by introducing *Rhizobium* both in saline and alkali soils [18]. Microbes play a dominant role in mobilization of immobile elements like phosphorus, zinc and copper in soils high in pH. These organisms release nutrients by organic acid or net excretion of H^+ or HCO_3^- ions. Nitrogen fixing algae like *Tolypothrix tenuis* and *Scytonema cincinnatum* can be used in waterlogged areas which fix nitrogen and are useful to improve yield of crops. In the saline and alkali soils organic contents are usually low. In these soils, repeated application of blue-green algae (BGA) bio-fertilizer, helps to bring down the level of soluble salts, pH and sodium content in exchange complex [19]. Hence, soil organic carbon content is not only maintained but also enriched too. Thus the microbes can be utilized to overcome the harmful effect of chemical degradation of soil and waterlogging which improves soil fertility.

Acknowledgements

Author acknowledges the contribution of Director, Central Soil Salinity Research Institute, Karnal, India for his encouragement during doing the study and research.

Author details

Shishir Raut
ICAR-Central Soil Salinity Research Institute, South 24 Parganas, West Bengal, India

*Address all correspondence to: shi_cssri94@yahoo.com

References

[1] Rao MSR. Land degradation problems and their management for sustainable production in India. In: Deb DL, editor. Natural Resources Management for Sustainable Agriculture and Environment. New Delhi: Angkor; 1994. pp. 125-142

[2] FAO. A provisional methodology for soil degradation assessment. Soils Bulletin, 84. Rome, Italy: FAO of United Nations

[3] Raheja PC. In: Boyko H, editor. Salinity and Aridity—New Approaches to Old Problems. Hague: Dr. W. Junk; 1966. pp. 43-127

[4] Chaudhary JS, Dhir RP. Journal of the Indian Society of Soil Science. 1985;**30**:342-347

[5] Kar A. Geology Review (India). 1989;**51**:148-159

[6] Joshi DC, Dhir RP. Arid Soil Research and Rehabilitation. 1990;**4**:141-148

[7] Ghose B. In: Proceedings of the Symposium on Problems of Indian Arid Zone. Jodhpur, India: CAZRI; 1971

[8] Singh NT. Land degradation and remedial measures. In: Deb DL, editor. Natural Resources Management for Sustainable Agriculture and Environment. New Delhi: Angkor; 1994. pp. 109-124

[9] Shishir R, Maji B, Burman D. Study of soil physicochemical parameters and organic matter humic components in relation to salinity in different landforms in a coastal soil of West Bengal. International Journal Environmental & Agricultural Science. 1998;**1**:1-5

[10] Kononova MM. Soil Organic Matter. Oxford: Pergamon Press; 1966. 456p

[11] Weil RR. Humic/fulvic acid ratios of some surface soils. American Journal of Alternative Agriculture. 1993;**8**:5-14

[12] Kononova MM, Nowakowski TZ, Greenwood GA. Soil Organic Matter—Its Nature, its Role in Soil Formation and in Soil Fertility. Oxford: Pergamon Press; 1961. 544p

[13] Dyke AV, Johnson PG, Grossl PR. Humic substances effect on moisture retention, nutrition, and colour of Intermountain West Putting Greens. USGA: Turfgrass and Environmental Research Online. 2009;**8**:1-9

[14] Singh R, Das DK. Wettability of soil under different plant covers. Journal of the Indian Society of Soil Science. 1992;**40**:39-43

[15] Kaushik BD, Subhashini D. Amelioration of salt affected soils with blue green algae. Proceedings of the Indian National Science Academy. 1985;**51**:386-389

[16] Subhashini D, Kaushik BD. Amelioration of salt affected soils with blue green algae. Phykos. 1984;**23**:273-277

[17] Kaushik BD. Reclamation potential of cyanobacteria in salt affected soils. Phykos. 1989;**28**:101-109

[18] AICPA. All India Coordinated Project on Algae. Technical Report. New Delhi: DST, GOI

[19] Dinesh K. Microbial tools for resource management. In: Deb DL, editor. Natural Resources Management for Sustainable Agriculture and Environment. New Delhi: Angkor; 1994. pp. 417-442

Section 2

The Role of Bacterial and Fungal Communities on Enhancing Nutrient Bioavailability

Chapter 6

Role of Fungi in Agriculture

Muthuraman Yuvaraj and Murugaragavan Ramasamy

Abstract

Fungi are a group of eukaryotic organisms and source of food, organic acids, alcohol, antibiotics, growth-promoting substances, enzymes, and amino acids. They include microorganisms like molds, yeasts, and mushrooms. They live on dead or living plants or animals' tissue. Fungi are very different from other living organisms; they are the primary decomposers of substances in the ecological system. Fungi are tremendous decomposer of organic waste material and most readily attack cellulose, lignins, gums, and other organic complex substances. Fungi can act also under a wide range of soil reaction from acidic to alkaline soil reactions. Fungi conjointly play a basic role in different physiological processes as well as mineral and water uptake, chemical change, stomatal movement, and biosynthesis of compounds termed biostimulants, auxins, lignan, and ethylene to enhance the flexibility of plants to ascertain and cope environmental stresses like drought, salinity, heat, cold, and significant metals.

Keywords: fungi, mycorrhiza, plant growth

1. Introduction

The microorganism was used from the very beginning of the civilization in the agriculture and industrial processes even before their existence was well known. Production of fermented beverages, bread and vinegar are traditional processers practiced from the time of early civilization. Recent advancement in our understanding about the genetics, physiology, and biochemistry of fungi, has led the exploitation of fungi for preparation of different agriculture and industrial products of economic importance. All the environmental factors influence the distribution of the fungal flora of soil [1, 2].

The primary functions of filamentous fungi in the soil are to degrade organic matter and help in soil aggregation. Besides this property, bound species of *Alternaria, genus Aspergillus, Cladosporium, Dematium, Gliocladium, Humicola* and *Metarhizium* manufacture substance like organic compounds in soil and therefore could also be necessary for the maintenance of soil organic matter. Plant growth regulators and chemical fertilizers have been used to increase crop production [3, 4]. Application of chemical fertilizers to crop plants negatively affects human health and environments. Recent studies have focused on identification of alternative methods to enhance plant productivity and protect the soil. Soil borne microbes can enter roots and establish their population in plants as endophytes, and many plant-associated fungi are well known for their capacity to promote plant growth; however, the relationship between these microbes and plants is still uncertain [5]. Microorganisms have the ability to produce phytohormones, solubilize insoluble phosphate and convert complex organic substances to simple forms. Endophytic fungi have also been shown to impart plants with tolerance to salt, drought, heat and diseases [6].

The four endophytic fungi (GM-1, GM-2, GM-3, and GM-4) were tested for their ability to improve soybean plant growth under salinity stress conditions. The seed germination and plant growth were higher in seeds pretreated with endophytic fungal cultures than their controls. The positive influence of fungi on plant growth was supported by gibberellins analysis of culture filtrate (CF), which showed wide diversity and various concentrations of Gibberellic acids [7].

Application of rhizospheric fungi is an effective and environmentally friendly method of improving plant growth and controlling many plant diseases. Three predominant fungi (PNF1, PNF2, and PNF3) isolated from the rhizospheric soil of peanut plants were screened for their growth-promoting efficiency on sesame seedlings. Among these isolates, PNF2 significantly increased the shoot length and fresh weight of seedlings compared with controls. Analysis of the fungal culture filtrate showed a higher concentration of indole acetic acid in PNF2 than in the other isolates [8].

The fungal associations with plants influence the primary and secondary metabolism of plants at all developmental stages. Photosynthesis is an important primary mechanism, and the main source of energy for plants. Its efficiency is related to photosynthetic pigments such as chlorophylls and carotenoids. Leaf chlorophyll a was increased in fungi-treated plants more so than in the controls [9].

2. Role of soil fungus

The fungi dominate in low pH or slightly acidic soils where soils tend to be undisturbed [10]. Fungi break down the organic residues so many alternative sorts of microbes will begin to decompose and method the residues into usable merchandise. Approximately 90% of all plants form symbiotic mycorrhizae fungi relationships by forming hyphae networks. Through mycorrhizae the plant obtains mainly phosphate and other minerals, such as zinc and copper, from the soil. The fungus obtains nutrients, such as sugars, from the plant root. This mutually beneficial relationship is called a mycorrhizae network [11].

Soil fungi can grow in a wide range of soil pH but their population is more under acidic conditions because of severe competition with bacteria at neutral pH. A majority of fungi are aerobic and prefer to grow at optimum soil moisture. The contribution of these organisms in biochemical transformation under excessive moisture is negligible [12].

The rhizosphere is a locality next to the basis dominated by soil microbes wherever several chemicals and organic chemistry methods occur. Soil fungi form up to 10–30% of the soil rhizosphere. The fungi ability to produce a wide variety of extracellular enzymes, they are able to break down all kinds of organic matter, decomposing soil components and thereby regulating the balance of carbon and nutrients for maintain soil health. This allows fungi to bridge gaps in the soil to transport nutrients relatively far distances back to the plants [13] (**Tables 1** and **2**).

Soil is a primary source of fungal growth, and is associated with the roots of all plant species. Fungi produce a wide range of bioactive metabolites, which can improve plant growth [14]. In addition, fungi supply inorganic nutrients to plants, such as ammonium, nitrate, and phosphate [15] and they are used as biofertilizers. Rhizosphere microorganisms can overcome competition with other soil factors and survive under variable environmental conditions [16].

Fungal species/strain	Plant type	Fungi-mediated response	Beneficial effects on plant species	References
AM fungi	Dead vegetation in soil	Degrade of dead organic	Nutrient mobilization	[43] Hodge et al. (2001)
Phanerochaete velutina	Wood	Decomposing wood	Phosphorus translocation	[44] Wells et al. (1998)
Pleurotus sp.	Wood	Wood decay	Nutrient mobilization	[45] Cohen et al. (2002)
Perisporiopsis lateritia	Leaves of *Hevea* sp.	Leaves decay	Nutrient mobilization	[46] Chaverri and Gazis (2010)
Navisporus floccosus	Wood	Wood decay	Nutrient mobilization	[47] Phillips et al. (2012)
M fungi	*Pinus taeda*	Decomposing organic matter	Carbon and nitrogen cycling	[48] Hoorman (2011)
AM fungi	*Vigna unguiculata*	Mineral uptake	Improved nutritional status	[49] Yaseen et al. (2011)
M fungi	*Allium cepa*	Plant growth	Improved nutritional status	[50] Albrechtova et al. (2012)
Trichoderma sp.	*Arabidopsis sp.*	Auxins dependent mechanism	Higher biomass production and increased lateral roots formation	[51] Contreras-Cornejo et al. (2009)
Trichoderma sp.	Agriculturally important crops	Biocontrol	Crop management	[52] Chalot and Brun (1998), [53] Harman and Mastouri (2010)
Ectomycorrhizal fungi	Higher plant species	Phenolic compounds degradation	Plant protection	[54] Ha (2010)
Ectomycorrhizal fungi and AM fungi	Agricultural crops	Stomatal physiology and water relation	Improved water potential status and increased photosynthesis rate	[55] Arnold and Engelbrecht (2007)

Table 1.
Soil-beneficial fungi on different physiological and catabolic processes in various host plant species.

Product	Microorganism used	Agriculture application
Gibberellins	*Fusarium moniliforme*	Plant growth hormone
Zearalenone	*Fusarium graminearum*	Growth promoter in cattle
DeVine	*Phytophthora palmivora*	Control of milkweed vine
Collego	*Colletotrichum sp*	Control of northern jointvetch
Chontral	*Chondrostereum purpureum*	Control of hardwoods
Rotstop	*Phanerochaete gigantea*	Control of butt rot of conifers

Table 2.
Agricultural application of fungi.

3. Ecological plant-microbe interactions

The microbes and plants along regulate several soil processes as well as the carbon cycle and nutrient utilization. Plant diversity and abundance might modification the complete soil scheme through the discharge of root exudates that attract or inhibit the expansion of specific organisms [17].

3.1 Economic advantage of fungi

- The saprophytic fungi of decay maintain the never-ending cycle of greenhouse emission that could be the most significant staple for plant chemical processes in nature. They additionally cause rot, decay, and decomposition of animal and plant remains emotional plant nutrients in an exceedingly type offered to inexperienced.

- There are types of fungi they serve to suppress fungi inflicting the sickness disease of the seedlings and thereby influence favorably the expansion of crops.

- Some fungi like *Empusa sepulchrasis, Metarhizium anisopliae,* and *Cordyceps melothac* can be used to control some insect pests. Others parasitic to some insects particularly, some spore-forming ones. The fungi spores sprayed on the crop cuss to regulate them. Colorado potato beetles, citrus rust mites, and spittle-bugs of insect cuss that may be controlled exploitation fungi. These types of fungi form loops on their mycelium which traps and strangle nematodes as the attempt to pass through. They later absorb nutrition from the nematodes.

3.2 Vesicular arbuscular mycorrhiza

Vesicular arbuscular mycorrhiza (VAM) fungi belong to the *Glomeromycota*. They are primitive fungi at the base of the tree for higher fungi (basidiomycetes). They turn out microscopic structures, or comparatively tiny sporocarps (truffle-like). Just over 200 species of these fungi are described, yet they are capable of forming mycorrhizal associations with the majority of plants. The word mycorrhiza is derived from the classical Greek word for "mushroom" and "root." In a mycorrhizal association, the underground mycellium is in contact with plant roots, but without causing any harm to the plant.

Mycorrhizal fungi accountable in the rising growth of host plant species because of raised nutrient uptake, production of growth-promoting substances and tolerance to drought, salinity and synergistic interactions with other beneficial microorganisms [18]. The soil conditions prevalent in sustainable agriculture are likely to be more favorable to AM fungi than are those under conventional agriculture [19]. The AM fungi are widely distributed in natural and agricultural environments and have been found associated with more than 80% of land plants, ferns, woody gymnosperms and angiosperms and grasses [20].

Arbuscular mycorrhiza fungi (AMF) are beneficial fungal organisms that share symbiotic association with many land plants. The arbuscular mycorrhiza fungi have the potential to improve soil characteristics, thereby promoting plant growth in normal and stressful environments [21]. The arbuscular mycorrhiza fungi colonization enhances plant growth [22] and changes the morphological, nutritional and physiological levels of plants to improve resistance against different abiotic stresses [23]. The arbuscular mycorrhiza fungi inoculation protects *Ocimum basilicum* against salinity stress by improving mineral uptake, chlorophyll synthesis and water use efficiency [24]. Tomato plants inoculated with arbuscular mycorrhiza fungi show an increase in the leaf area, nitrogen, potassium, calcium and phosphorous contents to enhance the plant growth rate compared to controls [25].

3.3 Edible fungi

Fungi can be used to produce material of nutritive value such as vitamins, amino acids, and lipids to make it more nutritious and palatable. Mushrooms are cultivated to yield fruit bodies directly consumed as food and yeast cells, mold mycelium is grown in fermenters to produce single-cell protein which may be used as food.

3.4 Plant response to AM Fungal inoculation

Soil phosphorus is a critical factor in plant response and responses are generally better under low phosphorus levels. Host genotypes and fungal strains seem to influence the response of plants to inoculation. The worldwide field experiment has provided evidence to show that under marginal P-deficiency soils lacking in effective AM fungal endophytes increase in yield of wheat, maize, barley, potatoes, and cowpea. Increased uptake of zinc has also been shown in AM fungus inoculated peach, maize, wheat and potato in zinc deficiency soils. The AM associations related to increased uptake of sulfur and calcium, improved water absorption and tolerance of plants to water stress in citrus and avocado seedlings have also been noticed. There are also reports of increased levels of cytokinins and chlorophyll by AM fungus- infected plants [26]. Therefore, many researchers were trying to use alternative approaches based on either manipulating or adding microorganisms to enhance plant protection against pathogens. The useful microorganisms (antagonistic bacteria) (e.g., bacteria genus visible radiation, *Bacilli subtilis*) and fungi (e.g., AMF, Trichoderma) contend with plant pathogens for nutrients and house, by manufacturing antibiotics, by parasitizing pathogens [27].

3.5 Exploitation of AM fungi for nutrient uptake and exchange

The fungi form a symbiotic association with roots of higher plants, facilitating uptake of plant nutrients, particularly of those which are less mobile this association is known as mycorrhizal association [28].

There are two types of mycorrhizal association (i) Ectotrophic mycorrhizae and (ii) Endomycorrhizae.

i. Ectotrophic mycorrhizae

Ectotrophic mycorrhizae, where the fungus forms a mantle or sheath around the root surface and where the mycelium develops intracellularly. The fungi which forms this types of association are species of *Boletus, Amenita*, etc.

ii. Endomycorrhizae

Endomycorrhizae, where the fungus develops intracellularly in the root without forming Hartig net. In this association the penetration of roots cells is characterized by the formation of terminal spherical structure called vesicular, which contain oil droplets and phosphorus. This type of mycorrhiza is called vesicular arbuscular mycorrhizae.

The management of AM fungi is very vital for organic and low-input agriculture systems wherever soil phosphorus is, in general, low, although all agroecosystems can benefit by promoting arbuscular mycorrhizae establishment. Some crops that poor at seeking out nutrients within the soil passionate about AM fungi for phosphorus uptake. For example, flax, which has poor chemotaxis ability, is highly dependent on AM-mediated phosphorus uptake at low and intermediate soil phosphorus concentrations. Proper management of AMF in the agroecosystems can improve the quality of the soil and the productivity of the land. Agricultural practices like reduced tillage, low phosphorus fertilizer usage and perennialized cropping systems promote functional mycorrhizal symbiosis [29].

3.6 Function of AM fungi in soil quality and phytoremediation

The use of arbuscular mycorrhizal fungi in ecological restoration comes (phytoremediation) has been shown to modify host plant institution on degraded soil and improve soil quality and health. There is evidence to suggest that this enhancement of soil aggregated stability is due to the production of a soil protein known as glomalin [30]. The arbuscular mycorrhizal fungi and is of agricultural significance particularly in the Phosphorus deficient soils where the where the phosphorus in the vesicle diffuses out into the cytoplasm and is taken up by the plant. Fungi belonging to the genera *Glomus, Endogene* form this association [31].

3.7 Role of AM fungi in salinity problem

The mycorrhizas can be used to help plants overcome extreme environmental conditions, such as saline environments [32] and several AM species have been found living in saline habitats [33]. According to some estimates, around 50% of plants living near shorelines possess mycorrhizal associations in their root systems [34]. Similarly, several species of AM were discovered in salt marsh plants [35]. Even in very saline sites reaching more than 150 dS/m of electrical conductivity, there are species of AM that can survive such hostile conditions [36].

There are different mechanisms by which AM fungi can help plants cope with salt stress. For example, they can enhance soil nutrient absorption by plants [37, 38] showed that the addition of AM fungi to lettuce and onion plants resulted in increased accumulation of phosphorus under conditions of salinity stress. Furthermore, AM can affect the ionic balance of plants, especially about Na^+ and Cl^- [39].

Furthermore, the addition of AM to tomato (*Lycopersicon esculentum*) under conditions of salinity improved anti-oxidant enzyme production, thus protecting cell membranes from damage. AM fungi can also improve the secretion of different types of hormones, one of them being abscisic acid. Mycorrhizal effects on hormones are important, as these hormones can enable plants to overcome many environmental stressed [40]. For example, inoculation of lettuce (*Lactuca sativa*) with Glomus intraradices induced enhanced levels of hormones in these plants under conditions of salinity stress and this, in turn, affected the regulation of stomatal closure. Salinity may also induce drought conditions for plants, so AM fungi may also help plants increase water uptake. The addition of mycorrhizas to leek (*Allium porrum*) increased the surface area of the roots, thereby increasing water absorption by the plants. The efficiency of water use in lettuce plants improved significantly with the addition of mycorrhizas under salt stress [41].

3.8 Potential of AM fungi in drought condition

Rice is mostly cultivated under rain-fed conditions. The yield can be severely reduced when the water supply is insufficient, therefore drought is one of the major

Product	Fungus	Target
Mycotal	*Verticillium lecanii*	Whitefly and thrips
Vertalec	*Verticillium lecanii*	Aphids
Biogreen	*Metarhizium anisopliae*	Scarab larvae on pasture
Cobican	*Metarhizium anisopliae*	Sugarcane spittle bug
Conidia	*Beauveria bassiana*	Coffee berry borer
Ostrinil	*Beauveria bassiana*	Corn borer
CornGuard	*Beauveria bassiana*	European corn borer

Table 3.
Mycoinsecticide.

constraints for rice production. Rice has its mechanisms to drought stress, and they are also assisted by living soil organisms. Arbuscular mycorrhizal (AM) fungi are among one of the soil microorganisms that may enhance drought resistance of rice. It assists plants in uptake water and nutrients. It also plays roles in regulating plant hormones, as well as stomatal behavior under drought stress. Apart from that, intercropping is likely contributing to the improvement of drought resistance and AM fungi activity. Intercropping can enhance AM fungi colonization and improve the root morphology of rice which beneficial for drought resistance. Thus, this analysis aims to achieve a lot of insight regarding the mutuality between AM fungi and rice beneath drought stress. The study will focus on the effects of AM fungi on the growth of rice, rice hormones, water potential and the contribution of AM fungi and intercropping on drought resistance of rice. The mycorrhizal development still strongly stimulated the improvement of plant growth and increased plant survival under drought stress. AMF had shown to reinforce drought tolerance in numerous plants [42].

3.9 Mycoinsecticides

The fungi have been utilized for controlling insect pests. The microbial control of insect pests emerged 100 years ago. Insect is infected by fungi through the body surface and this property is different from the infection caused by bacteria, viruses, and protozoa. Fungi attacking insect are called entomogenous. The conidia of the insect attacking fungi are attached to the insect integument where they germinate and the germ tubes penetrate in insect body under optimum temperature and humidity. The fungus proliferates in the insect body and the insect body gets covered with mycelia and conidia. The newly formed conidia are dispersed and cause subsequent infections and the cycle is continued (**Table 3**).

3.10 Myconematicides

Based on the nature of fungal biocontrol agents the nematopathogenic fungi are of three types, nematode, trapping fungi (*Arthrobotrys, Dactylella*), endoparasites (*Hirsutella, Meria*) and highly specific egg parasites (*Datylella*). The common and commercialized myconematicide are Royal 300 R (*Arthrobotrys robata*), Royal 350 R (*Arthrobotrys suporba*).

4. Conclusions

The increased absorption of available nutrients from soil as the fungus changes root morphology, which result in the larger root surface available for nutrient absorption. Fungal filaments also act as the absorption surface and increasing the nutrient availability by solubilizing insoluble nutrients like phosphorus, which thus become available to plant and increasing the nutrient mobility due to faster intracellular nutrient mobility and mobilizing nutrients from the soil mass not visited by the roots system but traversed by the mycorrhizal hyphae. The arbuscular mycorrhizal fungi protected plants by up-regulating the activity of antioxidant enzymes and osmolytes and by regulating the synthesis of phytohormones, which might possibly interconnect the various tolerance mechanisms for cumulative stress response. The prominent effect of arbuscular mycorrhizal fungi against salinity was proven to be due to a restriction in sodium uptake by roots and to the homeostasis of nutrient uptake.

Author details

Muthuraman Yuvaraj* and Murugaragavan Ramasamy
Department of Soil Science and Agricultural College, Adhiparasakthi Agricultural College, Tamil Nadu, India

*Address all correspondence to: yuvasoil@gmail.com

References

[1] Lavelle P, Spain AV. Soil Ecology and Chapter 3: Soil Organisms. New Delhi, India: Springer; 2005

[2] Magdoff F, Van H. Building Soils for Better Soil: Sustainable Soil Management, Chapter 4: The Living Soil. In: Sustainable Agriculture Network, Handbook Series Book 10. 3rd ed. Beltsville, Maryland: SARE Sustainable Agriculture Research & Education; 2009

[3] Islam KR. Lecture on Soil Physics, Personal Collection of R. Islam. Columbus, OH: The Ohio State University School of Environment and Natural Resources; 2008

[4] Metting FB. Structure and physiological ecology of soil microbial communities. n F.B. Metting, Jr. (Ed.), Soil Microbial Ecology, Marcel Dekker, Inc.: New York. 1993; pp. 3-25

[5] Sylvia DM, Hartel PG, Fuhrmann JJ, Zuberer DA. In: Sylva DM, editor. Principles and Applications of Soil Microbiology (2nd Ed.). Upper Saddle River, New Jersey: Pearson Prentice Hall; 2005

[6] Lowenfels J, Lewis W. Teaming with Microbes A gardener's Guide to the Soil Food Web, Chapter 3: Bacteria. Portland, Oregon: Timber Press; 2006

[7] Radhakrishnan R, Khan AL, Lee IJ. Endophytic fungal pre-treatments to seeds alleviate salinity stress effects in soybean plants. Journal of Microbiology. 2013;**51**(6):850-857

[8] Radhakrishnan R, Shim KB, Lee BW, Hwang CD, Pae SB, Park CH, et al. IAA producing *Penicillium* sp. NICS01 triggers plant growth and suppresses fusarium induced oxidative stress in sesame (*Sesamum indicum* L.). Journal of Microbiology and Biotechnology. 2013;**23**(6):856-863

[9] Radhakrishnan R, Kang SM, Baek IY, Lee IJ. Characterization of plant growth-promoting traits of *Penicillium* species against the effects of high soil salinity and root disease. Journal of Plant Interactions. 2014;**9**(1):754-762

[10] Sreenivasa MN, Bagyaraj DJ. Use of pesticides for mass production of vesicular arbuscular mycorrhizal inoculum. Plant and Soil. 1989;**119**:127-132

[11] Smith SE Read DJ. Mycorrhizal Symbiosis. 2nd ed. London, UK: Academic Press; 1997. p. 605

[12] Smith SE, Read DJ. Mycorrhizal Symbiosis. 3rd ed. London: Academic Press; 2008. p. 800

[13] Sturz AV, Carter MR, Johnston HW. A review of plant disease, pathogen interactions and microbial antagonism under conservation tillage in temperate humid agriculture. Soil and Tillage Research. 1997;**41**:169-189

[14] Waqas M, Khan AL, Lee IJ. Bioactive chemical constituents produced by endophytes and effects on rice plant growth. Journal of Plant Interactions. 2014;**9**:478-487

[15] Seastedt TR, Hobbs RJ, Suding KN. Management of novel ecosystems are novel approaches required? Frontiers in Ecology and the Environment. 2008;**6**:547-553

[16] Ferrara FIS, Oliveira ZM, Gonzales HHS, Floh EIS, Barbosa HR. Endophytic and rhizospheric enterobacteria isolated from sugar cane have different potentials for producing plant growth-promoting substances. Plant and Soil. 2012;**353**:409-417

[17] Gianinazzi S, Schuepp H, Barea JM, Haselwandter K. Mycorrhizal Technology in Agriculture: From Genes

to Biproducts. Switzerland: Springer; 2002

[18] Maeder P, Fliessbach A, Dubois D, Gunst L, Fried P, Niggli U. Soil fertility and biodiversity in organic farming. Science. 2002;**296**:1694-1697

[19] Gianinazzi S, Bosatka M. Inoculum of arbuscular mycorrhizal fungi for production systems: Science meets business. Canadian Journal of Botany. 2004;**82**:64-71

[20] Alqarawi AA, Abdallah EF, Hashem A. Alleviation of salt-induced adverse impact via mycorrhizal fungi in *Ephedra aphylla* Forssk. Journal of Plant Interactions. 2014;**9**(1):802-810

[21] Mo Y, Wang Y, Yang R, Zheng J, Liu C, Li H, et al. Regulation of plant growth, photosynthesis, antioxidation and osmosis by an arbuscular mycorrhizal fungus in watermelon seedlings under well-watered and drought conditions. Frontiers in Plant Science. 2016;7:644

[22] Hashem A, Abdaallah EF, Alqarawi AA, Aldubise A, Egamberdieva D. Arbuscular mycorrhizal fungi enhances salinity tolerance of *Panicum turgidum* Forssk by altering photosynthetic and antioxidant pathways. Journal of Plant Interactions. 2015;**10**(1):230-242

[23] Shekoofeh E, Sepideh H, Roya R. Role of mycorrhizal fungi and salicylic acid in salinity tolerance of *Ocimum basilicum* resistance to salinity. Journal of Biotechnology. 2012;**11**(9):2223-2235

[24] Balliu A, Sallaku G, Rewald B. AMF inoculation enhances growth and improves the nutrient uptake rates of transplanted, salt-stressed tomato seedlings. Sustainability. 2015;7:15967-15981

[25] Jones MD, Smith SE. Exploring functional definitions of mycorrhizas are mycorrhizas always mutualisms? Canadian Journal of Botany. 2004;**82**:1089-1109

[26] Landis FC, Gargas A, Givnish TJ. Relationships among arbuscular mycorrhizal fungi, vascular plants and environmental conditions in oak savannas. The New Phytologist. 2004;**164**:493-504

[27] Tarafdar JC, Marschner H. Efficiency of VAM hyphae in utilization of organic phosphorus by wheat plants. Soil Science & Plant Nutrition. 1994;**40**:593-600

[28] Clark RB, Zeto SK. Mineral acquisition by arbuscular mycorrhizal plants. Journal of Plant Nutrition. 2000;**23**:867-902

[29] Hildebrandt U, Janetta K, Ouziad F, Renne B, Nawrath K, Bothe H. Arbuscular mycorrhizal colonization of halophytes in central European salt marshes. Mycorrhiza. 2001;**10**:175-183

[30] Liu R, Wang F. Selection of appropriate host plants used in trap culture of arbuscular mycorrhizal fungi. Mycorrhiza. 2003;**13**:123-127

[31] Cooke JC, Lefor MW. The mycorrhizal status of selected plant species from Connecticut wetlands and transition zones. Restoration Ecology. 1998;**6**:214-222

[32] Aliasgharzadeh N, Rastin SN, Towfighi H, Alizadeh A. Occurrence of arbuscular mycorrhizal fungi in saline soils of the Tabriz Plain of Iran in relation to some physical and chemical properties of soil. Mycorrhiza. 2001;**11**:119-122

[33] Asghari HR, Marschner P, Smith SE, Smith FA. Growth response of *Atriplex nummularia* to inoculation with arbuscular mycorrhizal fungi at different salinity levels. Plant and Soil. 2005;**273**:245-256

[34] Cantrell IC, Linderman RG. Pre inoculation of lettuce and onion with VA mycorrhizal fungi reduces deleterious effects of soil salinity. Plant and Soil. 2001:269-281

[35] Giri B, Kapoor R, Mukerji KG. Improved tolerance of Acacia nilotica to salt stress by arbuscular mycorrhiza, *Glomus fasciculatum* may be partly related to elevated K/Na ratios in root and shoot tissues. Microbial Ecology. 2007;**54**:753-760

[36] Zhang YF, Wang P, Yang YF, Bi Q, Tian SY, Shi XW. Arbuscular mycorrhizal fungi improve reestablishment of *Leymus chinensis* in bare saline alkaline soil: Implication on vegetation restoration of extremely degraded land. Journal of Arid Enviroments. 2011;**75**:773-778

[37] Jahromi F, Aroca R, Porcel R, Ruiz-Lozano JM. Influence of salinity on the in vitro development of *Glomus intraradices* and on the in vivo physiological and molecular responses of mycorrhizal lettuce plants. Microbial Ecology. 2008;**55**:45-53

[38] Ruiz-Lozano JM, Azcon R, Gomez M. Alleviation of salt stress by arbuscular mycorrhizal *Glomus* species in *Lactuca sativa* plant. Physiologia Plantarum. 1996;**98**:767-772

[39] Garg N, Manchanda G. Role of arbuscular mycorrhizae in the alleviation of ionic, osmotic and oxidative stresses induced by salinity in *Cajanus cajan* (L.) Millsp. (pigeonpea). Journal of Agronomy and Crop Science. 2009;**195**:110-115

[40] Husband R, Herre EA, Turner SL. Molecular diversity of arbuscular mycorrhizal fungi and patterns of host association over time and space in a tropical forest. Molecular Ecology. 2002;**11**:2669-2678

[41] Johnson NC, Zak DR, Tilman D, Pfleger DL. Dynamics of vesicular arbuscular mycorrhizal fungi during old-field succession. Oecologia. 1991;**86**:49-58

[42] Berg G, Smalla K. Plant species and soil type cooperatively shape the structure and function of microbial communities in the rhizosphere. FEMS Microbiology Ecology. 2009;**68**:1-13

[43] Hodge A, Campbell CD, Fitter AH, et al. An arbuscular mycorrhizal fungus accelerates decomposition and acquires nitrogen directly from organic matter. Nature. 2001;**413**:297-299

[44] Wells JM, Boddy L, Donnelly DP. Wood decay and phosphorus translocation by the cord forming basidiomycete *Phanerochaete velutina*: The significance of local nutrient supply. The New Phytologist. 1998;**138**:607-617

[45] Cohen R, Persky L, Hadar Y. Biotechnological applications and potential of wood-degrading mushrooms of the genus Pleurotus. Applied Microbiology and Biotechnology. 2002;**58**:582-594

[46] Chaverri P, Gazis RO. *Perisporiopsis lateritia*, a new species on decaying leaves of *Hevea* spp. from the Amazon basin in Peru. Mycotaxon. 2010;**113**:163-169

[47] Phillips RP, Meier IC, Bernhardt ES, Grandy S, Wickings K, Finzi AC. Roots and fungi accelerate carbon and nitrogen cycling in forests exposed to elevated CO_2. Ecology Letters. 2012;**15**:1042-1049

[48] Hoorman JJ. The Role of Soil Fungus (report no. SAG-14-11). Columbus, Ohio, USA: Ohio State University; 2011

[49] Yaseen T, Burni T, Hussain F. Effect of Arbuscular mycorrhizal inoculation on nutrient uptake, growth and productivity of cowpea (*Vigna*

unguiculata) varieties. African Journal of Biotechnology. 2011;**10**: 8593-8598

[50] Albrechtova J, Latr A, Nedorost L, Pokluda R, Posta K, Vosatka M. Dual inoculation with mycorrhizal and saprotrophic fungi applicable in sustainable cultivation improves the yield and nutritive value of onion. Scientific World Journal. 2012:1-8

[51] Contreras-Cornejo HA, Macias-Rodriguez L, Cortes-Penagos C, Lopez-Bucio J. *Trichoderma virens*, a plant beneficial fungus, enhances biomass production and promotes lateral root growth through an auxin-dependent mechanism in Arabidopsis. American Society of Plant Biologists. 2009;**149**:1579-1592

[52] Chalot M, Brun A. Physiology of organic nitrogen acquisition by ectomycorrhizal fungi and ectomycorrhizas. FEMS Microbiology Reviews. 1998;**22**:21-44

[53] Harman G, Mastouri F. The role of Trichoderma in crop management systems. Phytopathology. 2010;**100**:16

[54] Ha TN. Using *Trichoderma* species for biological control of plant pathogens in Vietnam. Journal of ISSAAS. 2010;**16**:17-21

[55] Arnold AE, Engelbrecht BMJ. Fungal endophytes nearly double minimum leaf conductance in seedlings of a neotropical tree species. Journal of Tropical Ecology. 2007;**23**:369-372

Chapter 7

Arbuscular Mycorrhiza-Associated Rhizobacteria and Biocontrol of Soilborne Phytopathogens

Meenakshi Singh, Manjari Mishra,
Devendra Kumar Srivastava and Pradeep Kumar Singh

Abstract

The mutualistic symbiosis of most land plants with arbuscular mycorrhizal (AM) fungi has been shown to favor mineral and water nutrition and to increase resistance to abiotic and biotic stresses. The main mechanisms involved in the control of the disease symptoms and intraradical proliferation of soilborne phytopathogens are due to root colonization with AM fungi. The role of the rhizobacteria is shown to be specifically associated with extraradical network of the AM and mycorrhizosphere. The mycorrhizosphere can form a favorable environment for microorganisms which have potentiality to act antagonistic to pathogen abundance. It makes an additional advantage in identifying rhizobacteria from AM fungi structures or mycorrhizosphere, which often lead to the isolation of organisms having strong properties of antagonism on various soilborne pathogens. The ability of AM fungi to control soilborne diseases is mainly related to their capacity to stimulate the establishment of rhizobacteria against the favorable environment of pathogen within the mycorrhizosphere prior to the root infection. Recent advancement in scientific research has provided more clear picture in understanding the mechanisms involved in AM fungi/rhizobacteria interactions. Herein, this chapter includes the mechanisms of the AM fungi-mediated biocontrol, interactions between AM-associated bacteria and AM fungus extraradical network, AM-associated bacteria and biocontrol activities and unfavorable zone to pathogen development: the mycorrhizosphere.

Keywords: AM-associated bacteria (AMB), arbuscular mycorrhizal fungi, biocontrol, mycorrhizosphere, soilborne pathogens

1. Introduction

A majority of land plants in nature are growing symbiotically in relationship with AM fungi. This relationship is well established with the roots of these plants. Soil exploration by the external mycelium of AM fungi increases the nutrient absorptive root surface area and thus favors the host plant in access to nutrients and water [1, 2]. Moreover, as the largest component of the soil microbial biomass [3, 4], AM fungi form widespread mycelial networks within the soil atmosphere, and hyphae harbour important sites for interactions with other soilborne microorganisms. The constricted zone adjacent to soil-living roots is called the

IntechOpen

rhizosphere [5]. It is characterized by increased microbial activity and by a specific microbial community structure [6, 7]. Along with root-AM fungi associations, factors influencing the community structure and the biomass of soil microorganisms lead to the establishment of a zone called mycorrhizosphere [8–12]. The zone of soil influenced by only AM fungi is called mycosphere. In the mycorrhizosphere, AM fungi structures and various rhizobacteria (AM fungi-associated rhizobacteria or AMB, e.g. *Paenibacilli*, *Bacilli* and *Pseudomonas* spp.) are generally identified by classical culture-dependent methods [13, 14]. It includes phospholipid fatty acid analysis (PLFA) [15] and polymerase chain reaction-denaturing gradient gel electrophoresis (PCR-DGGE) [13, 16, 17] which reinforce the hypothesis that AM fungi structures constitute important nutrient-rich niches for soilborne microorganisms. *Glomeribacter gigasporarum* (a new taxon of *Burkholderiaceae*) was even described as a Gram-non-cultivable (obligatory) bacterial endosymbiont of spore vacuoles, mycelium and intraradical hyphae of *Gigaspora margarita* [18]. *Glomeribacter gigasporarum* described in detail shows to be widespread within *Gigasporaceae*; it transmitted vertically and contains nitrogen fixation genes [19–21], while in *Gigaspora margarita*, it has been suggested and observed that this AM fungus might fix nitrogen and then deliver it to the symbiotic plant through the associated bacterial population [22]. The effects of this on host plant physiology can be recognized in mycorrhizal root colonization because of the consequence of the activity of specifically AM fungi-associated rhizobacteria.

The beneficial effects of AM fungi on the host plant physiology, in the decrease of intraradical and mycorrhizosphere population and in the decrease of disease symptoms of soilborne pathogens were reported in many biological systems, probably due to synergistic mechanisms [23–25]. The use of chemical pesticides are now avoided and not advocated in fields due to its risks to human health and the environment, and thus the implementation of sustainable agriculture has become essential in crop industry. The perception of the mechanisms involved in the AM fungi-mediated biocontrol will allow to maximize the performance of management of such sustainable agroecosystems and thus authorize the use of AM fungi and its benefits [26]. The main mechanisms involved in the biological control of diseases induced by soilborne phytopathogens start after root colonization with AM fungi especially due to its association with rhizobacteria which constitutes major element for this biocontrol.

2. Mechanisms of the AM fungi-mediated biocontrol

Reduction in the detrimental effects of soilborne pathogens after root colonization with AM fungi was described a long time ago [27, 28] and has been observed on various fungi, stramenopiles, nematodes and bacteria [12, 29]. Carlsen et al. [30] reported the total check of infectivity caused by *Pythium ultimum* on clover plants cv. Sonja by using *Glomus mosseae* as a symbiotic relation partner. For the biological control of pathogen, AM fungus or AM fungi/plant taxa association, conditions of culture, level of root colonization, time of AM fungus or pathogen inoculation and harvest, the mechanisms hypothesized, etc. should be involved [12, 23, 24, 29, 31–35]. The disease symptoms induced by pathogens can systemically be reduced in non-mycorrhizal roots of plants grown in AM fungi-inoculated split-root systems [36]. Various hypotheses have been suggested in an endeavor to elucidate the AM fungi-mediated biocontrol of soilborne phytopathogens. The fact that pathogen-induced symptoms are systemically regulated by AM fungi colonization is related to the establishment of induced systemic resistance (ISR) [37]. ISR is a resistance mechanism induced or acquired in plants which were already undergone for pretreatment with a variety of organisms and compounds [e.g. superoxide dismutases (SOD) and peroxidases, pathogenesis-related type 1 proteins (PR-1 proteins)].

Further, higher concentrations of phenolic acids could be detectable in plants which are colonized with AM fungi species subjected for biocontrol activities. Accumulation of jasmonic acid involved in the rhizobacteria-mediated ISR in mycorrhizal roots could be related to the systemic pathogen biocontrol [38, 39]. Cordier et al. [40] identified local cell wall modifications (callose accumulation around arbuscule-containing cortical cells of tomato roots). The synthesis of constitutive and additional isoforms of defense-related enzymes (e.g. chininases, chitosanases, β-1,3-glucanases, peroxidases and SOD) has also been locally detected in mycorrhizal roots [41–43]. The level of production of these enzymes or flavonoids was reported to be unrelated to the capacity of biocontrol of the AM fungi species [30, 44]. The transcript profiling and real-time quantitative PCR used to explore the transcriptional changes triggered by AM fungus colonization revealed a complex pattern of local and systemic changes in gene expression in roots of *Medicago truncatula* [45], and transcripts for defense-related proteins were reported to expressed locally. Furthermore, increase in concentrations of defense-related compounds such as rosmarinic acid, caffeic acids, phenolics and essential oils has not been recorded in colonization with *Glomus mosseae* which was reported for its role in protecting basil plants against *Fusarium oxysporum* f. sp. *basilica.* It highlights and indicates the role of other possible mechanisms in the AM fungus-mediated biocontrol activity which differs to stimulation of systemic and localized plant defense mechanisms [46].

The most commonly documented response to AM fungi colonization is an increase in phosphorus nutrition to the host plants which subsequently imparts more dynamic and more resistant properties against pathogen invasion. However, AM fungi-mediated biocontrol is unrelated to the soil phosphorus (P) availability and to the phosphorus status in plant tissues, thus possibly more dependent on other mechanisms [46–49].

Arbuscular mycorrhizal fungi normally compete for space and nutrients with soilborne pathogens within the zone of mycorrhizosphere and the host roots. Larsen and Bodker [50], using signature fatty acid profiles, demonstrated the decrease in biomass and energy reserves of both *Glomus mosseae* and *Aphanomyces euteiches* co-occupying pea roots; however *Phytophthora nicotianae* and *Glomus mosseae* never reported to occupy simultaneously in the same tomato root tissues [40]. A reduction in the extent of mycorrhizal colonization by different plant pathogens has been reported [51–54] indicating the possible occurrence of competitive interactions. The AM fungus is often inoculated before the attack of pathogen in order to favor biocontrol efficiency [54]. However, *Fusarium solani* f. sp. *phaseoli* genomic DNA quantified using quantitative real-time PCR was significantly reduced not only in the mycorrhizosphere and mycosphere but also in the bulk soil of a compartmentalized soil-root system which was inoculated with *Glomus intraradices*, whereas the AM fungus genomic DNA was not significantly modified by the pathogens in the soil [55]. Reduction in *Fusarium solani* f. sp. *phaseoli* growth as well as decrease in root rot symptoms as a result of colonization with *Glomus intraradices* could not be attributed to the competition for resources and habitat between the two fungi but mostly to the biotic or abiotic characteristic factors of the established mycorrhizosphere.

The extraradical network formed by *Glomus intraradices* around the roots of the plants has been reported to show a decrease in the growth of nematodes (e.g. *Radopholus similis* and *Pratylenchus coffeae*) and conidial formation of *Fusarium oxysporum* f. sp. *chrysanthemi*. In vitro aseptic conditions and the above-stated negative impacts are not important to affect the developmental stages of all nematodes, and it is also unrelated to the mycelial or spore densities of AM fungus [56–58]. Additionally, in the presence of the AM fungi, significant increase in spore germination and hyphal growth by *Fusarium oxysporum* f. sp. *chrysanthemi* was also reported, and thus, direct inhibition of pathogen by AM fungi structures could not properly be explained for biocontrol [56].

In vitro results of impact studies of the exudates of extraradical AM fungi network or by the mycorrhizal roots on pathogens are in contradiction. Crude extracts from the extraradical network of *Glomus intraradices* is clearly reported for the reduced germination of conidia of *Fusarium oxysporum* f. sp. *chrysanthemi* [59]. Similarly, inhibition in sporulation of pathogen *Phytophthora fragariae* is reported with exudates of strawberry roots which were colonized by *Glomus etunicatum* and *Glomus monosporum* [60]. During the harvest, compared to the exudates of non-AM-inoculated tomato roots, the exudates from in vitro grown AM (*Glomus intraradices*)-inoculated roots were reported either repulsive or more attractive for the zoospores of *Phytophthora nicotianae* [61].

Another example can be seen in the exudates of tomato roots which are reported to double the microconidia germination of *Fusarium oxysporum* f. sp. *lycopersici* in the presence of AM fungi *Glomus mosseae* compared to the exudates from non-mycorrhizal roots [54, 62]. The direct impact of exudates from mycorrhizal plants in the AM fungus-mediated biocontrol activity can directly be measured in soil conditions by quantification of the capacity of root infection by the pathogen [63]. Application of root exudates of tomato plants which are colonized with *Glomus intraradices* or *Glomus mosseae* has not been reported for any positive impact on another tomato plant for the control of pathogen *Phytophthora nicotianae,* while direct inoculation of these AM fungi (i.e. *Glomus intraradices* or *Glomus mosseae*) significantly reduced or controlled the growth of pathogen *Phytophthora nicotianae* in these other tomato plants. Thus, it suggests that exudates from one's mycorrhizal plant will not directly or indirectly inhibit the capacity of pathogen intraradical proliferation on other plants.

From the above it is evident that none of the cited mechanisms is involved in the AM fungus-mediated biocontrol, but it has been shown to happen in every plant-fungi system. These mechanisms might act in synergistic way with each other, with one mechanism becoming preponderant depending on the environmental conditions and the plant cultivar-pathogen/AM fungus strain. However, the mechanism related to the capacity of interaction of AM fungi with other soil microorganisms can significantly be attributed as one of the main reasons involved in the control of soilborne diseases.

3. Interactions between AM-associated bacteria and AM fungus extraradical network

The bacterial communities associated with various AM fungal inoculum or spores have been reported to differ from one another based on their association as one found in mycorrhizal isolate and others largely encountered in the mycosphere [15]. The species assemblages of cultivable bacteria from surface-disinfected spores of *Glomus mosseae* and *Glomus intraradices* were influenced both by fungal and plant species where 'spore type' is the important factor. This specificity of interaction in AM fungal species is usually hypothesized to be related to spore size and surface roughness. Under sterile conditions the bacterial adherence to spores or hyphae of AM fungi was demonstrated to be species-specific or depends on bacterial isolate and the fungal vitality [64]. The association competence of rhizobacteria to AM fungal surfaces could be dependent on their ability to form biofilms [65].

The roots colonized with *Gigaspora margarita* and its extraradical hyphae demonstrate that extracellular polysaccharides are involved in the in vitro association of *Pseudomonas fluorescens* CHAO to these biological surfaces [66]. *Pseudomonas fluorescens* CHAO have the abilities to form light spots, while two mucoid mutants of this strain by increased production of acidic extracellular polysaccharides formed a large number of clusters on non-mycorrhizal carrot roots, and mutants of *Azospirillum brasilense* and *Rhizobium leguminosarum* affected in extracellular polysaccharide production were strongly impaired in the capacity to attach to mycorrhizal root [67]. Strains of *Burkholderia* on *Gigaspora decipiens* were able to colonize the interior of the spores, and it demonstrates that AM fungal colonization does not occur on AM surfaces only through the biofilm formation [68]. Saprophytic activity of the bacteria was also observed by scanning electron microscopy (SEM) observations of *Glomus geosporum* spores [69]. The growth of *Pseudomonas chlororaphis* was also stimulated in presence of crude extracts, containing AM fungus exudates and mycelial compounds of AM fungi from the extraradical network of in vitro grown *Glomus intraradices* [59].

Arbuscular mycorrhizal fungi can stimulate the growth of rhizobacteria by providing nutritional resource through the release of exudates. Exudates collected from tomato roots which were colonized by *Glomus fasciculatum* were reported to attract *Azotobacter chroococcum* and *Pseudomonas fluorescens* more strongly than those collected from non-colonized roots [70]. According to Toljander et al. [71], a bacterial community extracted from soil was significantly affected after 48 h when inoculated with exudates produced by AM fungus mycelia in comparison to a control composed of culture medium.

The reduction in exudation through defoliation of pea plants did not change the PCR-DGGE profile of rhizosphere bacteria, while missing and supplementary bands were observed from the rhizosphere of plants which were pre-colonized with *Glomus intraradices* [72]. PCR-DGGE analysis reported to show no effect on the bacterial community structure of tomato rhizosphere which was treated with pre-colonized (with *Glomus intraradices* or *G. mosseae*) root exudates however direct colonization of root with these AM fungi-induced significant changes [24]. The rhizobacterial community structure modification by AM fungal colonization is usually related poorly to exudate liberation by mycorrhizal roots or by the AM fungal mycelium, and importantly it may be dependant on their physical presence or on direct species-specific interactions [24]. It has been noticed that the impact of AM fungus colonization on other soil microorganisms is negative. The overall decrease of microbial activity described after root colonization with AM fungi has been proposed to be due to competition for substrates [73]. In association with cucumber, *Glomus intraradices* possess negative effect on the population of *Pseudomonas fluorescens* DF57. This negative effect was reported in both rhizosphere and in mycosphere [74].

4. AM-associated bacteria and biocontrol activities

Most of AM-associated bacteria (AMB) described so far in detail showed antagonistic characteristics towards soilborne pathogens or behaved as mycorrhization helper [16]. Similar studies have been performed by various researchers in aiming to identify AMB with biocontrol activities. A bacterial strain of *Paenibacillus* sp. B2 has been isolated from the mycorrhizosphere of *Glomus mosseae* and identified by phylogeny of its 16S rRNA gene sequence and analytical profile index (API) system. It has been found that it acts antagonistic to various soilborne pathogens under in vitro conditions and reduces necrosis in tomato roots (necrosis caused by *Phytophthora nicotianae*) [75]. This isolate (i.e. bacteria) displayed cellulolytic, proteolytic, chitinolytic and pectinolytic activities and was reported for antibiotic polymyxin B1 and two other polymyxin-like compounds [76–78]. Moreover, its presence resulted in disorganization of cell walls and/or cell contents of *Phytophthora nicotianae* and *Fusarium oxysporum* as observed in electron microscope. It also increases the root and shoot fresh weights of mycorrhized tomato plants and stimulated *Glomus mosseae* to colonize tomato roots [75].

Under compartmentalized growth system, Mansfeld-Giese et al. [78] identified *Paenibacillus polymyxa* and *P. macerans* from the three different regions, namely mycorrhizosphere, hyphosphere (root-free soil and sand compartments) and from a root-free sand compartment. It was found to be closely associated with *Glomus intraradices*. All *Paenibacilli* strains tested from these AM fungi influenced soil zones and helped in preventing pre-emergence damping-off (caused by *Pythium aphanidermatum*) [79]. Out of 18 cultivable isolates from surface-disinfected spores of *Glomus mosseae*, 14 isolates were identified. These identified isolates were mainly composed of *Bacillus simplex*, *B. niacini*, *B. drententis*, *Paenibacillus* spp. and *Methylobacterium* sp. which were reported to show antagonism to various soilborne pathogens (e.g. *Phytophthora nicotianae*, *Fusarium solani*, *Fusarium oxysporum*, etc.) [80]. Bacteria isolated from surface-decontaminated spores of *Glomus intraradices* and *Glomus mosseae* which were extracted from rhizospheres of *Festuca ovina* and *Leucanthemum vulgare* were classified within two phylogenetic clusters: one corresponding to *Proteobacteria* and the other corresponding to *Actinobacteria* and *Firmicutes* [14]. Under dual culture in vitro assays, bacteria from both clusters were reported antagonistic to *Rhizoctonia solani*. Further, selected bacteria, two isolates of *Stenotrophomonas maltophilia*, three isolates of *Pseudomonas* spp., one isolate each of *Bacillus subtilis* and *Arthro bacterilicis*, were reported to act as antagonistic to *Erwinia carotovora* var. *carotovora*, *Verticillium dahliae*, *Phytophthora infestans* and *Rhizoctonia solani*. In vitro studies revealed that these isolates are responsible for producing siderophores and proteases and thus decrease the weight of rotten potato tissues [81]. The ability of AM fungi to specifically harbor and then to stimulate rhizobacteria with biocontrol properties suggests that these bacteria can directly reduce pathogen development within the mycorrhizosphere and they can strongly contribute to the biocontrol of soilborne diseases.

5. Unfavorable zone to pathogen development: the mycorrhizosphere

The mycorrhizosphere has been hypothesized to comprise of favorable surroundings for the growth and development of microorganisms which works antagonistic to soilborne pathogens proliferation. Undeniably, co-culture of the non-mycorrhizal species (e.g. *Dianthus caryophyllus*) with the mycorrhizal species (e.g. *Tagetes patula*) in the presence of AM fungi (e.g. *Glomus intraradices*) clearly reduces the disease caused by Fusarium *oxysporum* f. sp. *dianthi* in the plant *Dianthus caryophyllus*. It occurs in a manner which differs in providing nutrition to plants and thus suggests a decline in the pathogen development within the mycorrhizosphere [82]. Moreover, a reduction in the number of infection loci in tomato roots (pre-colonized with *Glomus mosseae* and also inoculated with *Phytophthora nicotianae* zoospores) infers that the pathogen may be affected prior to root penetration in the mycorrhizosphere [83].

The mycorrhizosphere influenced by the rhizobacteria + AM fungus + root tripartite associations presents specific characteristics, in which individual factor influences the others' growth and health. Remarkably in the presence of glycoproteins such as glomalin, AM fungi favor the formation of aggregates which provide stable microsites favorable to root and microbe establishment [84, 85]. The AM fungi extraradical network also constitutes specific microsites which favor the growth of some bacteria. Among different plant growth-promoting rhizobacteria, P-solubilizing and N-fixing bacteria has been reported for more efficient synergistic interaction with AM fungi. Increased P and N availability to the plants promotes its growth and probably favors its capacity to counteract pathogen impact [11, 86–88].

Plant growth-promoting rhizobacteria can also display biocontrol properties and impact pathogen proliferation through direct liberation of toxic compounds or by competing for space and nutrients, reduction of Fe and Mn availability, modification of the plant hormone balance and stimulation of plant defense mechanisms [89, 90]. A synergistic or additive impact by dual inoculation of AM fungi with rhizobacteria in controlling pathogens reflects the dependence of biocontrol properties on the combinations of bacterial and fungal species used, nutritional status in soil and probably other environmental conditions [87].

Reduction in gall formation and nematode multiplication (which are usually responsible for causing root rot in chick pea) was significantly reported in the tomato plants when its roots were inoculated together with *Glomus intraradices* and bacteria *Pseudomonas striata* and *Rhizobium* sp. [91]. Similar positive reports have been recorded when dual inoculation of *Glomus mosseae* with *Pseudomonas fluorescens* was done [92]. Jaderlund et al. [93] reported the interactions of two plant growth-promoting rhizobacteria, namely, *Pseudomonas fluorescens* SBW25 and *Paenibacillus brasilensis* PB177, with AM fungi *Glomus mosseae* and *Glomus intraradices,* respectively; he investigated it on winter wheat which was infested with *Microdochium nivale* and concluded that this interactions are species-specific between fungi and bacteria. From the above and several other studies, it is clear that microbial antagonist to pathogens, and fungi-plant growth-promoting rhizobacteria, do not exert any negative effect against AM fungi [87]. Thus, such mycorrhization helper bacteria (MHB) are important in promoting mycorrhizal development and may even increase AM fungi impact on pathogens.

6. Conclusion

The competence of AM fungi to control disease symptoms and the intraradical and rhizosphere proliferation of soilborne pathogens is multifaceted and influenced by different mechanisms possibly acting in a synergetic way with each other. Among these mechanisms, the capacity of extraradical network of AM fungi to stimulate beneficial microorganisms is possibly a strongly responsible factor involved. Different bacteria with high capacities of antagonistic activities against several soilborne pathogens have been reported within AM fungal extraradical structures and in the mycorrhizosphere of several AM fungi species. The AM fungi-mediated biocontrol activities can not solely be due to the AM fungus function but also related strongly to the capacity of the AM fungi to constitute an environment which favors the establishment of rhizobacteria with potential biocontrol abilities.

Acknowledgements

PKS, and MS thank the SERB, Department of Science and Technology, Government of India, for awarding Fast Track Young Scientist and for the financial assistance.

Author details

Meenakshi Singh[1], Manjari Mishra[2], Devendra Kumar Srivastava[3] and Pradeep Kumar Singh[3*]

1 Department of Environmental Studies, Panjab University, Chandigarh, India

2 School of Life Sciences, Jawaharlal Nehru University, New Delhi, India

3 Department of Botany, Eternal University, Baru Sahib, HP, India

*Address all correspondence to: singhpk@hotmail.co.in

References

[1] Rhodes LH, Gerdemann JW. Phosphate uptake zones of mycorrhizal and non-mycorrhizal onions. New Phytologist. 1975;**75**:555-561

[2] Smith SE, Read DJ. Mycorrhizal Symbiosis. San Diego, London: Academic Press; 2008

[3] Kabir Z, Ohalloran IP, Fyles JW, Hamel C. Seasonal changes of arbuscular mycorrhizal fungi as affected by tillage practices and fertilization: Hyphal density and mycorrhizal root colonization. Plant and Soil. 1997;**192**:285-293

[4] Olsson PA, Thingstrup I, Jakobsen I, Baath F. Estimation of the biomass of arbuscular mycorrhizal fungi in a linseed field. Soil Biology and Biochemistry. 1999;**31**:1879-1887

[5] Hiltner L. Uber neuereer Fahrungen und probleme auf demgebiet der bodenbakteruiligie und unterbes ondererberuck sichtiguang der grundungung und brache (in German). Arbeiten der Deutschen Landwirtschaftlichen Gesellschaft. 1904;**98**:59-78

[6] Duineveld BM, Kowalchuk GA, Keijzer A, Veen JAV. Analysis of bacterial communities in the rhizosphere of *Chrysanthemum* via denaturing gradient gel electrophoresis of PCR-amplified 16S rRNA as well as DNA fragments coding for 16S rRNA. Applied and Environmental Microbiology. 2001;**67**:172-178

[7] Kim JS, Dungan RS, Kwon SW, Weon HY. The community composition of root-associated bacteria of the tomato plant. World Journal of Microbiology and Biotechnology. 2006;**22**:1267-1273

[8] Linderman RG. Mycorrhizal interactions with the rhizosphere microflora: The mycorrhizosphere effect. Phytopathology. 1988;**78**:366-371

[9] Marschner P, Crowley DE, Lieberei R. Arbuscular mycorrhizal infection changes the bacterial 16S rDNA community composition in the rhizosphere of maize. Mycorrhiza. 2001;**11**:297-302

[10] Marschner P, Timonen S. Interactions between plant species and mycorrhizal colonization on the bacterial community composition in the rhizosphere. Applied Soil Ecology. 2005;**28**:23-36

[11] Lioussanne L, Beauregard MS, Hamel C, Jolicoeur M, St-Arnaud M. Interactions between arbuscular mycorrhiza and soil microorganisms. In: Khasa D, Piche Y, Coughlan A, editors. Advances in Mycorrhizal Biotechnology: A Canadian Perspective. Ottawa ON: NRC Press; 2009

[12] Singh PK, Vyas D. Biocontrol of plant diseases and sustainable agriculture. Proceedings of the National Academy of Sciences, India Section B. 2009;**79**(II):110-128

[13] Lioussanne L. Roles des modifications de la microflore bacterienne et de lexudation racinaire de la tomatepar la symbiose mycorhizi ennedans le biocontrole sur le *Phytophthora nicotianae*. Doctoral thesis. University of Montreal, Montreal [In French]. 2007

[14] Bharadwaj DP, Lundquist PO, Persson P, Alstrom S. Evidence for specificity of cultivable bacteria associated with arbuscular mycorrhizal fungal spores (multitrophic interactions in the rhizosphere). FEMS Microbiology Ecology. 2008;**65**:310-322

[15] Rillig MC, Lutgen ER, Ramsey PW, Klironomos JN, Gannon JE. Microbiota

accompanying different arbuscular mycorrhizal fungal isolates influence soil aggregation. Pedobiologia. 2005;**49**:251-259

[16] Xavier LJC, Germida JJ. Bacteria associated with *Glomus clarum* spores influence mycorrhizal activity. Soil Biology and Biochemistry. 2003;**35**:471-478

[17] Roesti D, Ineichen K, Braissant O, Redecker D, Wiemken A, Aragno M. Bacteria associated with spores of the arbuscular mycorrhizal fungi *Glomus geosporum* and *Glomus constrictum*. Applied and Environmental Microbiology. 2005;**71**:6673-6679

[18] Bianciotto V, Bonfante P. Arbuscular mycorrhizal fungi: A specialised niche for rhizospheric and endocellular bacteria. Antonie Van Leeuwenhoek. 2002;**81**:365-371

[19] Bonfante P. Plants, mycorrhizal fungi and endobacteria: A dialog among cells and genomes. Biology Bulletin. 2003;**204**:215-220

[20] JargeatP,CosseauC,OlahB,JauneauA, Bonfante P, Batut J, et al. Isolation, free-living capacities, and genome structure of *Candidatusglomeribacter gigasporarum*, the endocellular bacterium of the mycorrhizal fungus *Gigaspora margarita*. Journal of Bacteriology. 2004;**186**:6876-6884

[21] Bianciotto V, Genre A, Jargeat P, Lumini E, Becard G, Bonfante P. Vertical transmission of endobacteria in the arbuscular mycorrhizal fungus *Gigaspora margarita* through generation of vegetative spores. Applied and Environmental Microbiology. 2004;**70**:3600-3608

[22] Minerdi D, Fani R, Gallo R, Boarino A, Bonfante P. Nitrogen fixation genes in an endosymbiotic *Burkholderia* strain. Applied and Environmental Microbiology. 2001;**67**:725-732

[23] St-Arnaud M, Hamel C, Vimard B, Caron M, Fortin JA. Altered growth of *Fusarium oxysporum f*. sp. chrysanthemi in an in vitro dual culture system with the vesicular arbuscular mycorrhizal fungus glomus intraradices growing on *Daucus carota* transformed roots. Mycorrhiza. 1995;**5**:431-438

[24] Lioussanne L, Jolicoeur M, St-Arnaud M. The effects of arbuscular mycorrhizal fungi, of root exudates from mycorrhizal plants and of the soilborne pathogen *Phytophthora nicotianae* on the bacterial community structure of tomato rhizosphere. Soil Biology and Biochemistry. 2009;**42**:473-483

[25] Singh M, Srivastava DK, Singh PK. Synergistic effect of arbuscular mycorrhizal fungi and fusarium oxysporum f. sp. lycopersici in intercropping with tomato. Indian Phytopathology (In Press). 2019

[26] Gosling P, Hodge A, Goodlass G, Bending GD. Arbuscular mycorrhizal fungi and organic farming. Agriculture, Ecosystems & Environment. 2006;**113**:17-35

[27] Gerdemann JW. Vesicular-arbuscular mycorrhiza and plant growth. Annual Review of Phytopathology. 1968;**6**:397-418

[28] Gerdemann JW. Vesicular-Arbuscular Mycorrhiza. Academic Press; 1974

[29] Whipps JM. Prospects and limitations for mycorrhizas in biocontrol of root pathogens. Canadian Journal of Botany. 2004;**82**:1198-1227

[30] Carlsen SCK, Understrup A, Fomsgaard IS, Mortensen AG, Ravnskov S. Flavonoids in roots of white clover: Interaction of arbuscular mycorrhizal fungi and a pathogenic fungus. Plant and Soil. 2008;**302**:33-43

[31] Harrier LA, Watson CA. The potential role of arbuscular mycorrhizal (AM) fungi in the bioprotection of plants against soil-borne pathogens in organic and/or other sustainable farming systems. Pest Management Science. 2004;**60**:149-157

[32] Avis TJ, Gravel V, Antoun H, Tweddell RJ. Multifaceted beneficial effects of rhizosphere microorganisms on plant health and productivity. Soil Biology and Biochemistry. 2008;**40**:1733-1740

[33] Vierheilig H, Steinkellner S, Khaosaad T, Garcia-Garrido GM. The biocontrol effect of mycorrhization on soilborne fungal pathogens and the auto regulation of AM symbiosis: One mechanism, two effects? In: Varma A, editor. Mycorrhiza: Genetics and Molecular Biology–Eco-Function–Biotechnology–Eco-Physiology–Structure and Systematic. Berlin, Heidelberg: Springer-Verlag; 2008. pp. 307-320

[34] Akhtar MS, Siddiqui MA. Arbuscular mycorrhizal fungi as potential bioprotectants against plant pathogens. In: Siddiqui ZA, Akhtar S, Futai K, editors. Mycorrhizae: Sustainable Agriculture and Forestry. Berlin, Heidelberg: Springer; 2009. pp. 61-98

[35] Lioussanne L, Jolicoeur M, St-Arnaud M. Role of root exudates and rhizosphere microflora in the arbuscular mycorrhizal fungi-mediated biocontrol of *Phytophthora nicotianae* in tomato. In: Varma A, Kharkwal AC, editors. Symbiotic Fungi: Principles and Practice. Berlin, Heidelberg: Springer-Verlag; 2009. pp. 141-158

[36] Khaosaad T, Garcia-Garrido JM, Steinkellner S, Vierheilig H. Take-all disease is systemically reduced in roots of mycorrhizal barley plants. Soil Biology and Biochemistry. 2007;**39**:727-734

[37] Pozo MJ, Azcon-Aguilar C. Unravelling mycorrhiza-induced resistance. Current Opinion in Plant Biology. 2007;**10**:393-398

[38] Hause B, Maier W, Miersch O, Kramell R, Strack D. Induction of jasmonate biosynthesis in arbuscular mycorrhizal barley roots. Plant Physiology. 2002;**130**:1213-1220

[39] Hause B, Mrosk C, Isayenkov S, Strack D. Jasmonates in arbuscular mycorrhizal interactions. Phytochemistry. 2007;**68**:101-110

[40] Cordier C, Gianinazzi S, Gianinazzi-Pearson V. Colonisation patterns of root tissues by *Phytophthora nicotianae* var. *Parasitica* related to reduced disease in mycorrhizal tomato. Plant and Soil. 1996;**185**:223-232

[41] Pozo MJ, Dumas-Gaudot E, Slezack S, Cordier C, Asselin A, Gianinazzi S, et al. Induction of new chitinase isoforms in tomato roots during interactions with *Glomus mosseae* and/or *Phytophthora nicotianae* var *parasitica*. Agronomie. 1996;**16**:689-697

[42] Pozo MJ, Azcon-Aguilar C, Dumas-Gaudot E, Barea JM. Chitosanase and chitinase activities in tomato roots during interactions with arbuscular mycorrhizal fungi or *Phytophthora parasitica*. Journal of Experimental Botany. 1998;**49**:1729-1739

[43] Pozo MJ, Azcon-Aguilar C, Dumas-Gaudot E, Barea JM. Beta-1,3-glucanase activities in tomato roots inoculated with arbuscular mycorrhizal fungi and/or *Phytophthora parasitica* and their possible involvement in bioprotection. Plant Science. 1999;**141**:149-157

[44] Pozo MJ, Cordier C, Dumas-Gaudot E, Gianinazzi S, Barea JM, Azcon-Aguilar C. Localized versus systemic effect of arbuscular

mycorrhizal fungi on defence responses to *Phytophthora* infection in tomato plants. Journal of Experimental Botany. 2002;**53**:525-534

[45] Liu JY, Maldonado-Mendoza I, Lopez-Meyer M, Cheung F, Town CD, Harrison MJ. Arbuscular mycorrhizal symbiosis is accompanied by local and systemic alterations in gene expression and an increase in disease resistance in the shoots. Plant Journal. 2007;**50**:529-544

[46] Toussaint JP, Kraml M, Nell M, Smith SE, Smith FA, Steinkellner S, et al. Effect of *Glomus mosseae* on concentrations of rosmarinic and caffeic acids and essential oil compounds in basil inoculated with *Fusarium oxysporum* f. sp. *basilici*. Plant Pathology. 2008;**57**:1109-1116

[47] Trotta A, Varese GC, Gnavi E, Fusconi A, Sampo S, Berta G. Interactions between the soilborne root pathogen *Phytophthora nicotianae* var. *Parasitica* and the arbuscular mycorrhizal fungus *Glomus mosseae* in tomato plants. Plant and Soil. 1996;**185**:199-209

[48] Yao MK, Tweddell RJ, Desilets H. Effect of two vesicular-arbuscular mycorrhizal fungi on the growth of micropropagated potato plantlets and on the extent of disease caused by *Rhizoctonia solani*. Mycorrhiza. 2002;**12**:235-242

[49] St-Arnaud M, Elsen A. Interaction of arbuscular mycorrhizal fungi with soil-borne pathogens and nonpathogenic rhizosphere micro-organisms. In: Declerck S, Strullu SG, Fortin JA, editors. *In vitro* Culture of Mycorrhizas. Berlin, Heidelberg: Springer-Verlag; 2005. pp. 217-231

[50] Larsen J, Bodker L. Interactions between pea root-inhabiting fungi examined using signature fatty acids. New Phytologist. 2001;**149**:487-493

[51] Davis RM, Menge JA. Influence of *Glomus fasciculatus* and soil phosphorus on *Phytophthora* root rot of citrus. Phytopathology. 1980;**70**:447-452

[52] Baath E, Hayman DS. Plant growth responses to vesicular-arbuscular mycorrhizae XIV. Interactions with *Verticillium* wilt on tomato plants. New Phytologist. 1983;**95**:419-426

[53] Krishna KR, Bagyaraj DJ. Interaction between *Glomus fasciculatum* and *Sclerotium rolfsii* in peanut. Canadian Journal of Botany. 1983;**61**:2349-2351

[54] Singh PK, Singh M, Vyas D. Effect of root exudates of mycorrhizal tomato plants on microconidia germination of *Fusarium oxysporum* f. sp. *lycopersici* than root exudates from non-mycorrhizal tomato plants. Archives of Phytopathology and Plant Protection. 2010;**43**:1495-1503

[55] Filion M, St-Arnaud M, Jabaji-Hare SH. Quantification of *Fusarium solani* f. sp. *phaseoli* in mycorrhizal bean plants and surrounding mycorrhizosphere soil using real-time polymerase chain reaction and direct isolations on selective media. Phytopathology. 2003;**93**:229-235

[56] St-Arnaud M, Hamel C, Vimard B, Caron M, Fortin JA. Inhibition of *Fusarium oxysporum* f. sp. *dianthi* in the non-VAM species *Dianthus caryophyllus* by co-culture with *Tagetes patula* companion plants colonized by *Glomus intraradices*. Canadian Journal of Botany. 1997;**75**:998-1005

[57] Elsen A, Declerck S, Waele DD. Effects of *Glomus intraradices* on the reproduction of the burrowing nematode (*Radopholus similis*) in dixenic culture. Mycorrhiza. 2001;**11**:49-51

[58] Elsen A, Declerck S, Waele DD. Use of root organ cultures to investigate the

interaction between *Glomus intraradices* and *Pratylenchus coffeae*. Applied and Environmental Microbiology. 2003;**69**:4308-4311

[59] Filion M, St-Arnaud M, Fortin JA. Direct interaction between the arbuscular mycorrhizal fungus *Glomus intraradices* and different rhizosphere microorganisms. New Phytologist. 1999;**141**:525-533

[60] Norman JR, Hooker JE. Sporulation of *Phytophthora fragariae* shows greater stimulation by exudates of non-mycorrhizal than by mycorrhizal strawberry roots. Mycological Research. 2000;**104**:1069-1073

[61] Lioussanne L, Jolicoeur M, St-Arnaud M. Mycorrhizal colonization with *Glomus intraradices* and development stage of transformed tomato roots significantly modify the chemotactic response of zoospores of the pathogen *Phytophthora nicotianae*. Soil Biology and Biochemistry. 2008;**40**:2217-2224

[62] Scheffknecht S, Mammerler R, Steinkellner S, Vierheilig H. Root exudates of mycorrhizal tomato plants exhibit a different effect on microconidia germination of *Fusarium oxysporum* f. sp. *lycopersici* than root exudates from non-mycorrhizal tomato plants. Mycorrhiza. 2006;**16**:365-370

[63] Lioussanne L, Jolicoeur M, St-Arnaud M. The growth of the soilborne pathogen *Phytophthora nicotianae* is reduced in tomato roots colonized with arbuscular mycorrhizal fungi but unaffected by application of root exudates collected from corresponding mycorrhizal plants. Mycorrhiza. 2009;**19**:443-448

[64] Toljander JF, Artursson V, Paul LR, Jansson JK, Finlay RD. Attachment of different soil bacteria to arbuscular mycorrhizal fungal extraradical hyphae is determined by hyphal vitality and fungal species. FEMS Microbiology Letters. 2006;**254**:34-40

[65] Rudrappa T, Biedrzycki ML, Bais HP. Causes and consequences of plant-associated biofilms. FEMS Microbiology Ecology. 2008;**64**:153-166

[66] Bianciotto V, Bandi C, Minerdi D, Sironi M, Tichy HV, Bonfante P. An obligately endosymbiotic mycorrhizal fungus itself harbors obligately intracellular bacteria. Applied and Environmental Microbiology. 1996;**62**:3005-3010

[67] Bianciotto V, Andreotti S, Balestrini R, Bonfante P, Perotto S. Extracellular polysaccharides are involved in the attachment of *Azospirillum brasilense* and *Rhizobium leguminosarum* to arbuscular mycorrhizal structures. European Journal of Histochemistry. 2001;**45**:39-49

[68] Levy A, Chang BJ, Abbott LK, Kuo J, Harnett G, Inglis TJJ. Invasion of spores of the arbuscular mycorrhizal fungus *Gigaspora decipiens* by *Burkholderia* spp. Applied and Environmental Microbiology. 2003;**69**:6250-6256

[69] Roesti D, Ineichen K, Braissant O, Redecker D, Wiemken A, Aragno M. Bacteria associated with spores of the arbuscular mycorrhizal fungi *Glomus geosporum* and *Glomus constrictum*. Applied and Environmental Microbiology. 2005;**71**:6673-6679

[70] Sood SG. Chemotactic response of plant-growth promoting bacteria towards roots of vesicular-arbuscular mycorrhizal tomato plants. FEMS Microbiology Ecology. 2003;**45**:219-227

[71] Toljander JF, Lindahl BD, Paul LR, Elfstrand M, Finlay RD. Influence of arbuscular mycorrhizal mycelial exudates on soil bacterial growth and community structure. FEMS Microbiology Ecology. 2007;**61**:295-304

[72] Vestergard M, Henry F, Rangel-Castro JI, Michelsen A, Prosser JI, Christensen S. Rhizosphere bacterial community composition responds to arbuscular mycorrhiza, but not to reductions in microbial activity induced by foliar cutting. FEMS Microbiology Ecology. 2008;**64**:78-89

[73] Raiesi F, Ghollarata M. Interactions between phosphorus availability and an AM fungus (*Glomus intraradices*) and their effects on soil microbial respiration, biomass and enzyme activities in a calcareous soil. Pedobiologia. 2006;**50**:413-425

[74] Ravnskov S, Nybroe O, Jakobsen I. Influence of an arbuscular mycorrhizal fungus on *Pseudomonas fluorescens* DF57 in rhizosphere and hyphosphere soil. New Phytologist. 1999;**142**:113-122

[75] Budi SW, van Tuinen D, Martinotti G, Gianinazzi S. Isolation from the *Sorghum bicolour* mycorrhizosphere of a bacterium compatible with arbuscular mycorrhiza development and antagonistic towards soilborne fungal pathogens. Applied and Environmental Microbiology. 1999;**65**:5148-5150

[76] Budi SW, van Tuinen D, Arnould C, Dumas-Gaudot E, Gianinazzi-Pearson V, Gianinazzi S. Hydrolytic enzyme activity of *Paenibacillus* sp. strain B2 and effects of the antagonistic bacterium on cell integrity of two soil-borne pathogenic fungi. Applied Soil Ecology. 2000;**15**:191-199

[77] Selim S, Negrel J, Govaerts C, Gianinazzi S, van Tuinen D. Isolation and partial characterization of antagonistic peptides produced by *Paenibacillus* sp. strain B2 isolated from the sorghum mycorrhizosphere. Applied and Environmental Microbiology. 2005;**71**:6501-6507

[78] Mansfeld-Giese K, Larsen J, Bodker L. Bacterial populations associated with mycelium of the arbuscular mycorrhizal fungus *Glomus intraradices*. FEMS Microbiology Ecology. 2002;**41**:133-140

[79] Li B, Ravnskov S,Xie G, Larsen J. Biocontrol of *Pythium* damping-off in cucumber by arbuscular mycorrhiza-associated bacteria from the genus *Paenibacillus*. BioControl. 2007;**52**:863-875

[80] Lioussane L. Review. The role of the arbuscular mycorrhiza-associated rhizobacteria in the biocontrol of soilborne phytopathogens. Spanish Journal of Agricultural Research. 2010;**8**(S1):S51-S61

[81] Bharadwaj DP, Lundquist PO, Alstrom S. Arbuscular mycorrhizal fungal spore-associated bacteria affect mycorrhizal colonization, plant growth and potato pathogens. Soil Biology and Biochemistry. 2008;**40**:2494-2501

[82] St-Arnaud M, Vujanovic V. Effect of the arbuscular mycorrhizal symbiosis on plant diseases and pests. In: Hamel C, Plenchette C, editors. Arbuscular Mycorrhizae in Crop Production. NY: Haworth's Food Products Press; 2007. pp. 67-122

[83] Vigo C, Norman JR, Hooker JE. Biocontrol of the pathogen *Phytophthora parasitica* by arbuscular mycorrhizal fungi is a consequence of effects on infection loci. Plant Pathology. 2000;**49**:509-514

[84] Rillig MC, Mummey DL. Mycorrhizas and soil structure. New Phytologist. 2006;**171**:41-53

[85] Singh PK, Singh M, Tripathi BN. Glomalin: an arbuscular mycorrhizal fungal soil protein. Protoplasma. 2013;**250**:663-669

[86] Johansson JF, Paul LR, Finlay RD. Microbial interactions in the mycorrhizosphere and their significance

for sustainable agriculture. FEMS Microbiology Ecology. 2004;**48**:1-13

[87] Barea JM, Pozo MJ, Azcon R, Azcon-Aguilar C. Microbial co-operation in the rhizosphere. Journal of Experimental Botany. 2005;**6**:1761-1778

[88] Artursson V, Finlay RD, Jansson JK. Interactions between arbuscular mycorrhizal fungi and bacteria and their potential for stimulating plant growth. Environmental Microbiology. 2006;**8**:1-10

[89] Nehl DB, Allen SJ, Brown JF. Deleterious rhizosphere bacteria: An integrating perspective. Applied Soil Ecology. 1997;**5**:1-20

[90] Bowen GD, Rovira AD. The rhizosphere and its management to improve plant growth. Advances in Agronomy. 1999;**66**:1-102

[91] Akhtar MS, Siddiqui ZA. Biocontrol of a root rot disease complex of chickpea by *Glomus intraradices*, *Rhizobium* sp. and *Pseudomonas striata*. Crop Protection. 2008;**27**:410-417

[92] Siddiqui ZA, Mahmood I. Effect of a plant growth promoting bacterium, an AM fungus and soil types on the morphometrics and reproduction of *Meloidogyne javanica* on tomato. Applied Soil Ecology. 1998;**8**:77-84

[93] Jaderlund L, Arthurson V, Granhall U, Jansson JK. Specific interactions between arbuscular mycorrhizal fungi and plant growth-promoting bacteria: as revealed by different combinations. FEMS Microbiology Letters. 2008;**287**:174-180

Chapter 8

Ectomycorrhizal Fungi as Biofertilizers in Forestry

José Alfonso Domínguez-Núñez and Ada S. Albanesi

Abstract

Ectomycorrhizal (ECM) fungi play a fundamental role in the nutrient cycle in terrestrial ecosystems, especially in forest systems. In this chapter, the value of ECM fungi is reviewed from a global framework, not only to increase the production of edible fruit bodies and biomass of plants but also for the regular practices of reforestation and restoration of ecosystems, with implicit applications in biofertilization, bioremediation, and control of soil pathogens. Ecological functions of the ECM fungi are briefly reviewed. The direct implications of the ECM fungi in forestry are described. To do so, its role as a biotechnological tool in forest nursery production is briefly analyzed, as well as the role of mycorrhizal helper bacteria (MHB). Subsequently, the direct role as biofertilizers of the ECM fungi in forest management is discussed: reforestation, plantation management, and ecosystem restoration. The importance of ECM fungi to increase the tolerance of plants against biotic or abiotic stresses is analyzed.

Keywords: forestry, ectomycorrhiza, restoration, sustainable development, nutrients

1. Introduction

It was Albert Bernhard Frank (1885), a forest pathologist, who for the first time introduced the term mycorrhiza. In the Greek language, "mykes" refers to fungus and "rhiza" refers to root. Since Frank's description of mycorrhizal association in the 1880s [1], a lot of work has been generated by different investigators as a consequence of which it is estimated that 86% of terrestrial plant species are benefited as they acquire their mineral nutrients via mycorrhizal roots [2]. These groups of fungi establish a symbiotic relationship with the roots of plants, called mycorrhizas. Frank established two large subdivisions of mycorrhizas, ecto- and endomycorrhizas. Ectomycorrhizal fungi form mantle and Hartig network of intercellular hyphae in the roots of forest species. Endomycorrhizas are classified as arbuscular mycorrhizas, ericoid mycorrhizas, arbutoid mycorrhizas, monotropoid mycorrhizas, ectendomycorrhizas, or orchid mycorrhizas [3]. The Arbuscular Mycorrhizal fungi (AM) form arbuscules and vesicles, they are more variable than ECM fungi since they form symbiosis with trees and herbaceous plants. Each of these categories is characterized by the invasion of plant root cells by fungal hyphae but differs in the nature of intracellular hyphal development [4, 5].

Ectomycorrhizal fungi are predominantly *Basidiomycetes*, some *Ascomycetes*, and a very few *Zygomycetes*. In these symbiotic structures, the Hartig network is the interface for the metabolic exchange between the fungus and the root. The mycorrhizal mantle is connected to the filaments of fungi that extend into the soil (extraradical mycelium), directly involved in the mobilization, absorption, and translocation of soil nutrients and water to the roots. Molecular clock analysis on the reconciled tree suggested that ECM fungi evolved far later than the appearance of the last common ancestor of brown and white rot fungi about 300 mya [6]. These results supported the long-standing hypothesis that ECM fungi evolved polyphyletically from multiple saprophytic species. More than 7000 species of fungi form ectomycorrhizas [7], many of them with important commercial trees such as poplar, birch, oak, pine, and spruce [8]. The reproductive structures (fruiting bodies) of the macromycetes are known as mushrooms when they grow in the soil and, like truffles, when they grow underground.

The community of mycorrhizal fungi can be determinant in the structure of the plant community [9]. Therefore, the identification of the mycobiont partner and its functional structure [10] are fundamental to understand the ecological importance of this symbiotic relationship. ECM fungal diversity studies were initially based on studies of fruiting bodies and, more recently, on the direct identification of ectomycorrhizal morphoanatomical characters [11]. Despite recent advances in the use of molecular techniques, there are still many advantages associated with classical methods for studying ECM fungal diversity. For the recognition of fungal relationship and type of mycorrhizal association is advantageous over molecular method [7]. Sometimes morphoanatomical-based taxonomy is not well supported by molecular taxonomy. To overcome such discrepancy, the combined approach of morphoanatomical and molecular characterization of ectomycorrhizas in combination with phylogeny was applied [12].

Most of the cultivated species of edible fungi are saprophytes, and only some of them are ECM fungi [13]. The tickets (*Boletus edulis*), the chanterelles (*Cantharellus* spp.), the matsutake mushroom (*Tricholoma matsutake*), and the truffle (many species of the *Tuber* genus) are some ECM fungi for which the crop has been studied [14–16]. The black truffle or Périgord, *Tuber melanosporum*, is widely grown, while other species of ECM mushrooms have not yet been cultivated, including fungi porcini (*Boletus edulis* S.) and the high-priced Italian fungus, white truffles (*Tuber magnatum*).

2. Ecological functions of ECM fungi

In different forest ecosystems, ECM fungi have been reported to play an important role in seedling survival, establishment, and growth [3, 17, 18]. Researches have confirmed that ECM fungi play a key role in terrestrial ecosystems as drivers of global carbon and nutrient cycles [19].

Some of these traditionally known functions of the ECM fungi on the ecosystem are:

ECM fungi increase the water and nutrient supply plant, extending the volume of land accessible to the plants.

Different fungal species (drought-sensitive hydrophilic or drought-tolerant hydrophobic) can have different effects on hydraulic redistribution patterns [20]. The mechanisms to enhance the acquisition of P by tree mycorrhizal roots are the extension of extramatrical mycorrhizal hyphae, the increase of inorganic P transfer, the increase of inorganic P transporters in the fungus/soil interface, the mobilization of organic P (labile) by emission of phosphatases, and the mobilization of mineral insoluble P by the emission of organic acids (LMWOAs) [21, 22].

The mechanisms of improvement in nitrogen (N) absorption would be the intervention in the mineral N cycle (NH_4^+, NO_3^-) and the assimilation of organic N (by emitting proteases, chitinases, and others) [23, 24].

Currently, recent advances in the knowledge of nutrient translocation processes in the fungus-plant and fungus-soil interaction are especially interesting, in particular the priority role of transporters of P, N, and C [25]. The inorganic P and mineral or organic forms of N, such as NH_4^+, NO_3^-, and amino acids (AA), are absorbed by transporters specialized located in the fungal membrane in the extraradical mycelium. NH_3/NH_4^+ and inorganic P (from polyphosphates) are imported from the symbiotic interface to the cells of the plant through selective transporters. Transporters of hexoses import carbon of plant origin into the fungus. The nutritional strategies seem to be different between symbiotic and pathogenic fungi, for example, in the translocation of C. Even different transport strategies have been found between ECM symbionts *Ascomycota* and *Basidiomycota*. The understanding of the different systems of transporters or nutrient channels involved both at the level of the extraradical mycelium and at the level of the symbiotic interface will clarify in the future the processes of nutrition in the plant-fungus and fungus-soil interaction. Also, looking at the fungal factors in the establishment of the symbiotic relationship, chitin-related molecules seem to be shared by pathogenic and arbuscular mycorrhizal fungi, opening the question of whether they could also function in signaling in ectomycorrhizal symbioses [26].

On the other hand, the ECM colonization of the root can provide protection against soil pathogens [27]. Also, the non-nutritive benefits to plants due to changes in water relations, the level of phytohormones, the assimilation of carbon, etc. have already been verified [3]. The carbon is transferred through the ECM fungal mycelium that connects different species of plants. This can reduce competition among plants and contribute to the stability and diversity of ecosystems [28]. The extraradical mycelium of the ECM fungi provides a direct pathway for the translocation of photosynthesized carbon to microsites in the soil and a large surface area for interaction with other microorganisms [29, 30]. Recently, Hupperts et al. [31] proposed two competing models to explain carbon mobilization by ectomycorrhizal fungi. "Saprotrophy model", where decreased allocation of carbon may induce saprotrophic behaviour in ectomycorrhizal fungi, resulting in the decomposition of organic matter to mobilize carbon and second, "nutrient acquisition model", where decomposition may instead be driven by the acquisition of nutrients locked within soil organic matter compounds. Moreover, epigeous and hypogeal sporocarps of ECM fungi are important food sources for placental and marsupial mammals [32]. The ectomycorrhizal roots, the mycelium, and the fruiting bodies of the fungi are important as food sources and habitats for invertebrates [33]. The hyphal networks produced by ECM fungi significantly alter and improve the structure of the soil [34]. In a global way, the ECM fungi improve the plant tolerance to (biotic and abiotic) environmental stresses.

3. Applications: ECM fungi to forestry

Much of our understanding of the functions of ECM fungi has come from research directed toward practical application in forestry. Some of the most common criteria considered for the selection of a most valued species or strain of ECM fungi (some of them implicit in others) are the abiotic criteria: climatic conditions such as temperature, insolation, and humidity; improvement of soil properties, such as texture and permeability; abiotic soil stress mitigation; soil contamination mitigation; soil metal mobilization; or nutrient cycling. There may also be criteria regarding the host, such as the plant/fungus specificity, the improvement of plant health, or the increase in the biomass of the plant. Finally, there are criteria regarding the fungus, such as abundance, effectiveness, propagules competitiveness, fungus growth rate, or edibility. Other criteria may be the conservation of native biodiversity, the functioning of the ecosystem, human health, food, nutraceutical value, etc. [30, 35].

3.1 ECM fungi in forest nurseries

Since the late 1950s, mycorrhizal fungi were utilized as biofertilizers to promote plant growth, because of their ability to increase the plant uptake of P, N, mineral nutrients, and water [36–38]. The idea of inoculating ECM fungi on seedlings in plant nurseries was developed by Fortin [39]. Vozzo and Hacskaylo [40] while working on ECM in the United States experimentally demonstrated that field survival and growth of tree seedlings with specific potential ECM enhance the performance of seedlings and contribute to the proper functioning of forest ecosystems.

Although successful inoculation of tree seedlings (already planted) in the field has been known, nursery inoculation is more common. Seedlings inoculated in the nursery can establish a healthy ECM system before planting. The challenge in the controlled synthesis of the ectomycorrhizal symbiosis is to produce a quality mycorrhizal plant, only colonized by the desired fungus. Accurate identification of the inoculum used and avoiding contamination during the growth of the inoculated plants are essential parts of the production process to avoid the introduction of unwanted species and to avoid the mixing of their genetic material with indigenous species [41]. The appropriate selection of suitable plant-host species is essential for the success of mycorrhization [42]. Relatively fast-growing fungi are generally preferred for inoculation because of their short incubation period. Unfortunately, many otherwise desirable ECM fungi grow slowly. According to Marx [43], fresh cultures are preferred to cultures repeatedly transferred and stored for several years. He further suggests passing important fungus cultures through a host inoculation and mycorrhiza formation followed by re-isolation, every few years to maintain mycorrhiza-forming capacity. Moreover, fungi, which produce large hyphal stands of rhizomorphs in the culture of the soil, may be superior in soil exploration and mineral uptake to those which lack rhizomorphic growth. On the other hand, the fruiting of the ECM fungi species is not based solely on the mycorrhizal state of the seedlings. After planting, in addition to the presence of indigenous competitors, the biotic and physicochemical characteristics of the soil also influence the persistence and spread of the cultivated fungus [44]. The type of ECM material used for inoculation can affect the success of a mycorrhizal inoculation program. In addition to remaining viable during storage and transport, the inoculant must also maintain its infectivity for several months after its introduction [45].

There are three main sources of fungal inoculum: soil, spores, and mycelium.

Initially, the soil or humus collected from the mycorrhizal plantation area was frequently used. Its main disadvantage is the lack of control of ECM species in the soil or of microorganisms and harmful germs. Another problem with this type of inoculant is that large amounts of soil are required to inoculate nursery plants. This method is widely used in developing countries, although it is currently discarded in mycorrhization programs. Also, planting mycorrhizal "nurse" seedlings or incorporating chopped roots of ECM hosts into nursery beds as a source of fungi for neighboring young seedlings has been successful [46].

Other sources of inoculum are the spores of fruit bodies collected in the field. The main advantages are that the spores do not require the extension of the aseptic culture and that the spore inoculum is not heavy [47]. Most of the recent research has been with *Pisolithus tinctorius*. Inoculation with spores of *Rhizopogon* species also appears promising. Abundant *Rhizopogon* mycorrhizas formed on seedlings produced from the coated seed of *Pinus radiata* D. Don with basidiospores of *Rhizopogon luteolus* [5]. However, it has three main drawbacks: (a) significant quantities of fruiting bodies are required and may not be available each year, (b) the success of the inoculation is highly dependent on the viability of the spores, and (c) the lack of genetic definition. Freeze-drying and storage at a low temperature in the dark is helpful to maintain its viability. The spores can be mixed with physical supports before the soil inoculation; suspended in water and soaked in the soil; sprinkled, sprayed or pelleted, and emitted to the ground; and encapsulated or coated on the seeds, and they can be embedded in hydrocolloid chips [47].

The most appropriate inoculum is the use of hyphae in a solid or liquid medium or substrate. Hyphae are cultivated mainly from sterile parts of fruiting bodies, less frequently from mycorrhiza due to their low (approx. 5–20%) success rate [48] and rarely from sclerotia [49] or sexual spores [50]. It is considered the most appropriate method since it allows the selection of particular strains of a fungus previously tested for its ability to promote the growth of plants [43]. Many species do grow well in culture, e.g., most species of *Suillus*, *Hebeloma*, *Laccaria*, *Amanita*, *Rhizopogon*, and *Pisolithus* genus. Liquid substrates have the advantage over solids because they are easily mixed and produce more uniform conditions for crop growth, but the risk of bacterial contamination and costs are higher [45]. On the other hand, the main advantages of the solid medium [51] are the reduction of bacterial contamination due to the lower water content, the low costs of the equipment, and the simplified design of the bioreactors. The main drawbacks of the use of mycelial inocula are that several species of ECM fungi are difficult to grow under laboratory conditions, or growth is very slow (due to the absence of their symbiont), and it is not always easy to produce large amounts of inoculum viable for large-scale nursery inoculation programs. Some advances have been made using mycelium encapsulated in "beads" of calcium alginate (e.g., [52]), but they have to be refrigerated. Inoculant beads can remain viable for several months under refrigeration, although the results vary between fungal species. For several species, the mycelial inoculum has been tested with trees of economic interest. This technique has great potential for the inoculation of seedlings in reforestation programs. For example, Rossi et al. [45] designed a bioreactor with the capacity to produce inoculum for 300 000 seedlings, enough to reforest 200 hectares. Based on a global demand of 3.0 billion cubic meters of wood, an estimated 4.3 tons of mycelium would be needed to inoculate 12 billion seedlings (5 g of dry mycelium per plant [45]). An advantage of alginate gel is the possibility of preparing a multimicrobial inoculant.

3.1.1 Ectomycorrhizal helper bacteria

The concept of "mycorrhiza helper bacteria" (MHB) was introduced in a "Tansley Review": Helper Bacteria—a new dimension of mycorrhizal symbiosis [53], which has led to new research in the plant-fungus model system, as for the meaning of these bacteria that promote the formation of mycorrhizas and cause many physiological effects of mutualistic interaction. In general, the ability of some microorganisms to influence the formation and functioning of the symbiosis is known, through activities of various kinds such as the activation of infective propagules of the fungus in presymbiotic stages [54], facilitating the formation of entry points in the root [55] and increase of the growth rate [56]. The MHB improve mycorrhiza formation, although the same MHB can benefit mycorrhization for certain fungi and be negative for others [57]. The above reflects the fungal specificity by isolate, which exemplifies the genetic distance between isolates of different origin.

Among the mechanisms presented by the MHB are:

a. Promotion of the establishment of the symbiosis by stimulation of the mycelial extension. The germination of spores and mycelial growth are improved by the production of growth factors [58].

b. Increased contact and colonization root-fungus: increase in the number of lateral roots, mediated by the production of phytohormones [59] and the improvement of radical colonization by induction of flavonoid production [60].

c. Reduction of the impact of adverse environmental factors on the mycelium of the mycorrhizal fungus. Bacteria can detoxify soils, restoring their conductivity, similarly freeing them from contamination generated by heavy metals [61], and reducing the concentrations of phenolic antagonist compounds produced by the same mycorrhizal fungi [62]. The rhizospheric microorganisms also have an effect on the growth of the plants, reaching a synergistic effect, where the presence of the microfungus and the other microorganism produces an increase in the growth, vigor, and protection of the plant [63]. These effects are based on activities such as the acquisition of nutrients, inhibition of the growth of pathogenic fungi [64], and improvement of the root ramification [65]. In recent years, a potential capacity of bacteria associated with ectomycorrhizas to fix atmospheric nitrogen has been suggested [66]. Several studies suggest a real possibility that the bacteria present in mycorrhizal tissues contribute to the nutritional needs of both the fungus (ascocarp development) and consequently the plants, by providing them with available nitrogen derived from atmospheric nitrogen (N_2).

MHB belong to a wide range of genera (*Burkholderia*, *Paenibacillus* [67]; *Pseudomonas*, *Bacillus* [68]; *Streptomyces* [69]). However, the molecular mechanisms by which MHB induce the growth of ECM fungi are not well described. Recently, changes in expression of genes involved in the development of certain ECM fungi have been studied at the molecular level in confrontations with MHB [70–73].

Research in mycorrhizas should, therefore, strive toward an improved understanding of the functional and molecular mechanisms involved in interactions in the mycorrhizosphere, in order to develop ad hoc biotechnology that allows the application of optimized combinations of microorganisms as effective inoculators within sustainable systems of plant production [74].

3.1.2 Polymicrobial formulations

A polymicrobial formulation containing a diverse mixture of beneficial rhizosphere microorganisms with multiple functionalities is attractive because combining different classes of soil organisms can take advantage of multiple plant growth-promoting mechanisms and could be applied to multiple crops [75–79]. A key concept in constructing effective polymicrobial multifunctional formulations is the selection and use of a right combination of rhizosphere bacteria and fungi that are mutually compatible, have complementary functionalities, effectively colonize the rhizosphere of the crop(s) of interest, and bring about a synergistic promotion of growth and yield of crop(s) [75, 80–82]. It is to be expected that well-designed multifunctional formulations such as the one described would be a welcome addition to the fast-growing inoculant enterprises worldwide. Such an inoculant is also expected to be eco-friendly and suitable for organic farming and other integrated production systems, where synthetic fertilizer inputs are not allowed or restricted by law. However, construction of such complex formulations is technically demanding [83].

Ectomycorrhizal fungi exhibit synergistic interactions with other plant-beneficial organisms such as symbiotic N_2-fixers. For example, ectomycorrhizal symbiosis enhanced the efficiency of inoculation of two *Bradyrhizobium* strains on the growth of legumes [84]. It is also of interest that similar synergies were seen when AM fungus (*Glomus mosseae*), ECM fungus (*Pisolithus tinctorius*), and *Bradyrhizobium* sp. were used together to inoculate *Acacia nilotica*; enhancement of N_2 fixation, growth, and dry biomass were observed when all three organisms were present [85, 86].

Also, using plant growth-promoting microorganism (PGPM) strains that form stable and effective biofilms could be a strategy for producing commercially viable inoculant formulations [78, 87]. A majority of plant-associated bacteria found on roots and in the soil are found to form biofilms [88]. Bacterial, fungal, and bacteria/fungal biofilms were suggested as possible inoculants. This is a novel and interesting idea, but to what extent this approach would be practiced remains to be seen [83].

3.2 Application of ECM fungi in forest management: restoration of ecosystems

The inoculation of ECM fungi can be done with the objective of producing edible carpophores but also because of its considerable value in forest management; in particular, they have had great importance in reforestation programs where it was expected that the quality and economic productivity of the plantations would increase [89]. The success of the plantations with mycorrhizal seedlings from the nursery depends on their ability to quickly access the nutrients and water available within the soil matrix [90]. The relationships between the various native edible ECM fungi have been, until relatively recently, insufficiently considered in the strategies of forest management [91].

In ectomycorrhizal plantations (productive or conservation reforestations), a consequence of the recognition of the advantages of fungal diversity in ecosystems will be an increase in the refusal to introduce potentially dominant species in mixed communities. On the other hand, unfortunately, it seems that many of those fungi selected for optimal colonization in the nursery have been poor competitors in the field, especially when the planting sites contained indigenous populations of mycorrhizal fungi. There are several possible explanations for the inoculation failure (from the nursery) to produce beneficial effects in the planting sites. Probably, among the most important of these is the inability of inoculum introduced to persist in the roots of the plant after the transfer of the nursery to the field. The soil conditions experienced in the nursery and with the plant growing in a container are very different from those of most of the planting sites; in addition, the raising, storage, and transport of seedlings can reduce the vigor of fine roots and their fungal associates. Species such as *Pisolithus tinctorius* (15 sub spp.), in circumstances such as degraded environments, with absence or scarcity of autochthonous mycorrhizal populations, have achieved the greatest success in inoculation programs [92]. In the case of edible ECM fungi, such as *Tuber melanosporum* (black truffle), the establishment of mycorrhizal plantations has always aimed at the production of carpophores, leaving aside the contribution of ecological functions of the symbiosis (in the plant, in the soil, and, in general, in the ecosystem) [93]. The example of mycorrhizal plantations for truffle production has been generally successful [94], obtaining productions from 6 to 7 years of implantation.

In the restoration of ecosystems, the biofertilization, the bioremediation, and the biocontrol of soil pathogens are prominent roles of the ectomycorrhizal fungi. Degraded ecosystems are the result of a wide range of characteristics and factors related to unfavorable land management or industrial activities. Environmental degradation of the soil is increasing worldwide at an alarming rate due to erosion, acidity, salinization, compaction, depletion of organic matter, and water scarcity. On the contrary, in a healthy ecosystem, there is a balanced microbiota of the soil, in such a way that the potential of pathogenic and mycorrhizal fungi coexists in apparent harmony. Ectomycorrhizal fungi can survive in extreme habitats with high or low temperature [95, 96], salt and metal concentration [97, 98], drought [99], and other circumstances related to the degradation of the ecosystem. The importance of ECM fungi in the balance of the ecosystem can be enormous, since they can be used to increase the tolerance of plants against biotic or abiotic stresses, especially their capacity to fix heavy metals or to degrade a wide variety of persistent organic compounds; to interact with soil bacteria; to attack fungi, bacteria, and pathogenic nematodes; and to improve the vegetative growth and the nutritional status of its symbiont plant.

It has been documented by several authors that mycorrhizal fungi improve the disease resistance of their host plant primarily by direct competition, enhanced or altered plant growth, nutrition and morphology, induced resistance, and development of antagonist microbiota. Direct competition or inhibition is reported to be due to the production and release of antibiotics and physical sheathing by the mantle of ECM [27, 100–102]. For example, ECM fungi have been shown to protect trees from *Phytophthora cinnamomi* infection along with supporting their survival and growth in comparison to non-mycorrhizal seedlings [35, 101, 102]. Thus, ECM fungi can also be used as a fungicide in nursery plantations for better growth, survival, and establishment of seedlings.

Under drought stress, ectomycorrhizal symbiosis has been documented to possess a remarkable capacity to the uptake of water and alter hydraulic properties of plant roots by altering both apoplastic and symplastic pathways and by their impact on plant aquaporins (AQPs) [103–106]. A symbiosis between plants and ECM fungi has been documented to help plants to cope with salt stress [97, 107–109]. Li et al. [110] reported that there is ECM fungus-mediated remodeling of ion flux which helps to maintain K+/Na+ homeostasis by increasing the release of Ca2+. Also, ECM fungi have been reported to change the plant phytohormone balance during salt stress [111, 112]. Research efforts are still in progress to select new pioneer symbiotic couples for land reforestation [113].

Till date, most studies have indicated that ECM plants accumulate less metal inside their tissue and grow better than non-mycorrhizal plants when exposed to heavy metal stress [114–118]. Also, Meharg and Cairney [119] revised potential ways in which ectomycorrhizal fungi might support rhizosphere remediation of persistent organic pollutants (POPs). Krupa and Kozdrój [120] documented the importance of mycorrhizal fungi in forming an efficient biological barrier for checking the movement of heavy metals into the host tissues. Recently, the importance of LMW organic acids and metal chelating agents (such as siderophores) from ECM fungi in the fixation of metal ions and their transmission or not to the root of the host plant has been described [121]. The cellular mechanisms involved in detoxification of heavy metals by mycorrhizal fungi include biosorption of metals to fungal cell wall, chelation of metal ion in the cytosol by compounds such as glutathione and metallothioneins, metal exclusion mechanisms in metal-tolerant ECM fungi, and the compartmentation of metals in the vacuole, where metal ions are probably complexed in a chemically inactive form [98, 118, 122, 123].

4. Conclusions

The ectomycorrhizal fungi are predominantly *Basidiomycetes* and *Ascomycetes*, which establish a symbiotic relationship with the roots of forest plants, and these are directly involved in the mobilization, absorption, and translocation of soil nutrients and water to the roots. Most of the known cultivated species of edible fungi are saprophytes, and some of them are ectomycorrhizal fungi, but there is a promising potential in the study and knowledge of new species of ECM fungi as potential wild collected edible mushrooms. ECM fungi play a key role in terrestrial ecosystems as drivers of global carbon and nutrient cycles; in the fungus-plant interface, the role of C and nutrient transporters seems a priority. Research in ectomycorrhizal fungi should focus on better understanding the functional and molecular mechanisms involved in fungus-plant and fungus-soil interactions. For decades, our understanding of the functioning of ectomycorrhizal fungi has allowed us their application in the forest area. In the nursery, the inoculation of ECM fungi is a more common method to produce ectomycorrhizal forest seedlings, and the mycelial inoculation has great potential in reforestation programs. We should aim to find the appropriate technology for the commercial techniques of multiplication and large-scale inoculation of the mycorrhizal inoculum and the application of optimized combinations of plant-microorganisms (e.g., MHB, PGPB) adopted under well-defined environmental and soil conditions. The role of ECM fungi as biofertilizers in bioremediation or biocontrol in plantations, reforestation, and environmental restoration has been fundamental up to now, and its importance in the balance of the ecosystem can be enormous, increasing the tolerance of plants against biotic and abiotic stress. The application of ectomycorrhizal fungi in current environmental problems as the oaks or pines decline, or the phytoremediation of contaminated soils, seems promising. Research is still underway to select new pioneer symbiotic relationships for land restoration and reforestation.

Author details

José Alfonso Domínguez-Núñez[1]* and Ada S. Albanesi[2]

1 E.T.S.I de Montes, Forestal y del Medio Natural, Universidad Politécnica de Madrid, Ciudad Universitaria, Madrid, Spain

2 Departamento de Microbiología, Facultad de Agronomía y Agroindustrias, Universidad Nacional Santiago del Estero, Santiago del Estero, Argentina

*Address all correspondence to: josealfonso.dominguez@upm.es

References

[1] Frank B. Ueber die auf Wurzelsymbiose beruhende Ernährung gewiser Bäume durch unterirdishe Pilze. Berichte der Deutschen Botanischen Gesellschaft. 1885;**3**:128-145

[2] Brundrett MC. Mycorrhizal associations and other means of nutrition of vascular plants: Understanding the global diversity of host plants by resolving conflicting information and developing reliable means of diagnosis. Plant and Soil. 2009;**320**:37-77. DOI: 10.1007/s11104-008-9877-9

[3] Smith SE, Read DJ. Mycorrhizal Symbiosis. 3rd ed. London: Academic Press; 2008. p. 787

[4] Peterson RL, Massicotte HB, Melville LH. Mycorrhizas: Anatomy and Cell Biology. Wallingford, Oxon: CABI Publishing, CAB International; 2004

[5] Sharma R. Ectomycorrhizal mushrooms: Their diversity, ecology and practical applications. In: Varma A, Prasad R, Tuteja N, editors. Mycorrhiza - Function, Diversity, State of the Art. Switzerland: Springer International Publishing; 2017. pp. 99-131

[6] Marcel G, van der Heijden A, Martin FM, Selosse MA, Sanders IR. Mycorrhizal ecology and evolution: The past, the present, and the future. New Phytologist. 2015;**205**:1406-1423. DOI: 10.1111/nph.13288

[7] Rinaldi AC, Comandini O, Kuyper TW. Ectomycorrhizal fungal diversity: Separating the wheat from the chaff. Fungal Diversity. 2008;**33**:1-45

[8] Wiensczyk AM, Gamiet D, Durall DM, Jones MD, Simard SW. Ectomycorrhizas and forestry in British Columbia: A summary of current research and conservation strategies. B.C. Journal of Ecosystems and Management. 2002;**2**:1-20

[9] Fitter AH. Darkens visible: Reflections on underground ecology. Journal of Ecology. 2005;**93**:231-243

[10] Agerer R. Exploration types of ectomycorrhizas. A proposal to classify ectomycorrhizal mycelial systems according to their patterns of differentiation and putative ecological importance. Mycorrhiza. 2001;**11**:107-114

[11] Horton TR, Bruns TD. The molecular revolution in ectomycorrhizal ecology: Peeking into the black-box. Molecular Ecology. 2001;**10**:1855-1871

[12] Mrak T, Kühdorf K, Grebenc T, Štraus I, Münzenberger B, Kraigher H. Scleroderma areolatum ectomycorrhiza on Fagus sylvatica L. Mycorrhiza. 2017;**27**:283-293. DOI: 10.1007/s00572-016-0748-6

[13] Savoie JM, Largeteau ML. Production of edible mushrooms in forests: Trends in development of a mycosilviculture. Applied Microbiology and Biotechnology. 2011;**89**:971-979

[14] Chang ST, Hayes WA. The Biology and Cultivation of Edible Mushrooms. New York: Academic; 1978

[15] Chevalier G. The truffle cultivation in France: Assessment of the situation after 25 years of intensive use of mycorrhizal seedlings. In: Proceedings of the first International Meeting on Ecology, Physiology and Cultivation of Edible Mycorrhizal Mushrooms. 3-4 July; Uppsala, Sweden: Swedish University of Agricultural Sciences; 1998

[16] Bencivenga M. Ecology and cultivation of Tuber magnatum Pico. In: Proceedings of the first International Meeting on Ecology, Physiology and Cultivation of Edible Mycorrhizal Mushrooms. 3-4 July; Uppsala, Sweden:

Swedish University of Agricultural Sciences; 1998

[17] Sebastiana M, Martins J, Figueiredo A, Monteiro F, Sardans J, Peñuelas J, et al. Oak protein profile alterations upon root colonization by an ectomycorrhizal fungus. Mycorrhiza. 2017;**27**:109-128. DOI: 10.1007/s00572-016-0734-z

[18] Tedersoo L, May TW, Smith ME. Ectomycorrhizal lifestyle in fungi: Global diversity, distribution, and evolution of phylogenetic lineages. Mycorrhiza. 2010;**20**:217-263. DOI: 10.1007/s00572-009-0274-x

[19] Kumar J, Atri NS. Studies on ectomycorrhiza: An appraisal. The Botanical Review. 2018;**84**:108-155. DOI: 10.1007/s12229-017-9196-z

[20] Prieto I, Roldán A, Huygens D, Alguacil MM, Navarro-Cano JA, Querejeta JI. Species-specific roles of ectomycorrhizal fungi in facilitating interplant transfer of hydraulically redistributed water between *Pinus halepensis* saplings and seedlings. Plant and Soil. 2016;**406**:15-27

[21] Harley JL, Smith SE. Mycorrhizal Symbiosis. Toronto: Academic Press; 1983

[22] Plassard C, Dell B. Phosphorus nutrition of mycorrhizal trees. Tree Physiology. 2010;**30**(9):1129-1139. DOI: 10.1093/treephys/tpq063

[23] Harley JL. The significance of mycorrhiza. Mycological Research. 1989;**92**:129-139

[24] Chalot M, Plassard C. Ectomycorrhiza and nitrogen provision to the host tree. In: Polacco JC, Todd CD, editors. Ecological Aspects of Nitrogen Metabolism in Plants. England: Wiley-Blackwell; 2011. pp. 69-94

[25] Bonfante P, Genre A. Mechanisms underlying beneficial plant-fungus interactions in mycorrhizal symbiosis. Nature Communications. 2010;**27**:1-48. DOI: 10.1038/ncomms1046

[26] Bonfante P. The future has roots in the past: The ideas and scientists that shaped mycorrhizal research. New Phytologist. 2018;**220**:982-995

[27] Duchesne LC, Peterson RL, Ellis BE. Pine root exudate stimulates antibiotic synthesis by the ectomycorrhizal fungi Paxillus involutus. New Phytologist. 1988;**108**:471-476. DOI: 10.1111/j.1469-8137.1988.tb04188.x

[28] Simard SW, Jones MD, Durall DM, Perry DA, Myrold DD, Molina R. Reciprocal transfer of carbon isotopes between ectomycorrhizal *Betula papyrifera* and *Pseudotsuga menziesii*. New Phytologist. 1997;**137**(3):529-542

[29] Sun Y-P, Unestam T, Lucas SD, Johanson KJ, Kenne L, Finlay RD. Exudation-reabsorption in mycorrhizal fungi, the dynamic interface for interaction with soil and other microorganisms. Mycorrhiza. 1999;**9**:137-144

[30] Suz L, Azul AM, Pino-Bodas R, Martín MP. Ectomycorrhizal fungi in biotechnology: Present and future perspectives. In: Kumar A, Prasad R, editors. Environment and Biotechnology. Lambert Academic Publishing. AG & Co; 2012. pp. 472-542

[31] Hupperts SF, Karst J, Pritsch K, Landhäusser SM. Host phenology and potential saprotrophism of ectomycorrhizal fungi in the boreal forest. Functional Ecology. 2017;**31**:116-126. DOI: 10.1111/1365-2435.12695

[32] Claridge AW, May TW. Mycophagy by Australian mammals. Australian Journal of Ecology. 1994;**19**:251-275

[33] Fogel R, Peck SB. Ecological studies of hypogeous fungi. I Coleoptera

associated with sporocarps. Mycologia. 1975;**67**:741-747

[34] Griffiths RP, Caldwell BA. Mycorrhizal mat communities in forest soils. In: Read DJ, Lewis DH, Fitter AH, Alexander IJ, editors. Mycorrhizas in Ecosystems. Wallingford, UK: CAB International; 1992. pp. 98-105

[35] Azul AM, Nunes J, Ferreira I, Coelho AS, Verissimo P, Trovao J, et al. Valuing native ectomycorrhizal fungi as a mediterranean forestry component for sustainable and innovative solutions. Botany-Botanique. 2014;**92**(2):161-171

[36] Feldmann F, Hutter I, Schneider C. Best production practice of arbuscular mycorrhizal inoculum. In: Varma A, Kharkwal AC, editors. Symbiotic Fungi: Principles and Practice. Vol. 18. Berlin Heidelberg: Springer; 2009. pp. 319-336

[37] Koide R, Mosse B. A history of research on arbuscular mycorrhiza. Mycorrhiza. 2004;**14**:145-163

[38] Miransari M. Soil microbes and plant fertilization. Applied Microbiology and Biotechnology. 2011;**92**:875-885

[39] Fortin JA. Synthesis of mycorrhizas on explants of the root hypocotyls of *Pinus sylvestris* L. Canadian Journal of Botany. 1966;**44**:1087-1092. DOI: 10.1139/b66-116

[40] Vozzo JA, Hacskaylo E. Inoculation of *Pinus caribaea* with ectomycorrhizal fungi in Puerto Rico. Forest Science. 1971;**17**:239-245

[41] Murat C, Martin F. Sex and truffles: First evidence of Périgord black truffle outcrosses. The New Phytologist. 2008;**180**:260-263

[42] Olivier JM. Progress in the cultivation of truffles. In: Van Griensven LJLD, editor. Science and Cultivation of Edible Fungi. Rotterdam: A. Balkema; 2000. pp. 937-942

[43] Marx DH. Ectomycorrhizal fungus inoculations: A tool for improving forestation practices. In: Mikola P, editor. Tropical Mycorrhiza Research. London: Oxford University Press; 1980. pp. 13-71

[44] Hortal S, Pera J, Parladé J. Field persistence of the edible ectomycorrhizal fungus *Lactarius deliciosus*: Effect of inoculation strain, initial colonization level, and site characteristics. Mycorrhiza. 2009;**19**:167-177

[45] Rossi MJ, Furigo A, Oliveira VL. Inoculant production of ectomycorrhizal fungi by solid and submerged fermentations. Food Technology and Biotechnology. 2007;**45**:277-286

[46] Sim M-Y, Eom A-H. Effects of ectomycorrhizal fungi on growth of seedlings of *Pinus densiflora*. Mycobiology. 2006;**34**:191-195

[47] Marx DH, Cordell CE. The use of specific ectomycorrhizas to improve artificial forestation practices. In: Whipps JM, Lumsden RD, editors. Biotechnology of Fungi for Improving Plant Growth: Symposium of the British. Cambridge: Cambridge University Press; 1989. pp. 1-25

[48] Molina R, Palmer JG. Isolation, maintenance, and pure culture manipulation of ectomycorrhizal fungi. In: Schenck NC, editor. Methods and Principles of Mycorrhizal Research. St Paul, Minnesota, USA: The American Phytopathological Society; 1982. pp. 115-129

[49] Trappe JM. Studies on *Cenococcum graniforme*. An efficient method for isolation from sclerotia. Canadian Journal of Botany. 1969;**47**:1389-1390

[50] Fries N, Birraux D. Spore germination in *Hebeloma* stimulated by living plant roots. Experimentia. 1980;**36**:1056-1057

[51] Cannel E, Moo-Young M. Solid-state fermentation systems. Process Biochemistry. 1980;**15**:24-28

[52] Le Tacon F, Jung G, Mugnier J, Michelot P. Efficiency in a forest nursery of an inoculant of an ectomycorrhizal fungus produced in a fermentor and entrapped in polymetric gels. Annals of Forest Science. 1983;**40**:165-176

[53] Garbaye J. Helper bacteria: A new dimension to the mycorrhizal symbiosis. New Phytologist. 1994;**128**(2):197-210

[54] Azcón-Aguilar C, Barea JM. Arbuscular mycorrhizas and biological control of soil-borne plant pathogens - an overview of the mechanisms involved. Mycorrhiza. 1996;**6**:457-464

[55] Linderman RG. Mycorrhizal interactions with the Rhizosphere Microflora - the rhizosphere effect. Phytopathology. 1988;**78**:366-371

[56] Carpenter-Boggs L, Loynachan TE, Stahl PD. Spore germination of *Gigaspora margarita* stimulated by volatiles of soil-isolated actinomycetes. Soil Biology and Biochemistry. 1995;**27**:1445-1451

[57] Garbaye J, Duponnois R. Specificity and function of mycorrhization helper bacteria (MHB) associated with the *Pseudotsuga menziesii-Laccaria laccata* symbiosis. Symbiosis. 1992;**14**:335-344

[58] Keller S, Schneider K, Rd S. Structure elucidation of auxofuran, a metabolite involved in stimulating growth of fly agaric, produced by the mycorrhiza helper bacterium Streptomyces AcH 505. Journal of Antibiotics. 2006;**59**:801-803

[59] Bending GD, Poole EJ, Whipps JM, Read DJ. Characterisation of bacteria from *Pinus sylvestris-Suillus luteus* mycorrhizas and their effects on root-fungus interactions and plant growth. FEMS Microbiology Ecology. 2002;**39**:219-227

[60] Xie ZP, Staehelin C, Vierheilig H, Iemkena W, Jabbouri S, Broughton WJ, et al. Rhizobial nodulation factors stimulate mycorrhizal colonization of nodulating and nonnodulating soybeans. Plant Physiology. 1995;**108**:1519-1525

[61] Brulé C, Frey-Klett P, Pierrat JC, Courier S, Gérard F, Lemoine MC, et al. Survival in the soil of the ectomycorrhizal fungus *Laccaria bicolor* and effect of a mycorrhiza helper *Pseudomonas fluorescens*. Soil Biology and Biochemistry. 2001;**33**:1683-1694

[62] Duponnois R, Garbaye J. Some mechanisms involved in growth stimulation of ectomycorrhizal fungi by bacteria. Canadian Journal of Botany. 1990;**68**:2148-2152

[63] Domínguez JA, Martin A, Anriquez A, Albanesi A. The combined effects of *Pseudomonas fluorescens* and *Tuber melanosporum* on the quality of *Pinus halepensis* seedlings. Mycorrhiza. 2012;**22**(6):429-436

[64] Budi SW, Van Tuinen D, Martinotti MG, Gianiazzi S. Isolation from the *Sorghum bicolour* mycorrhizosphere of a bacterium compatible with arbuscular mycorrhiza development and antagonistic towards soil-borne fungal pathogens. Applied and Environmental Microbiology. 1999;**65**:148-150

[65] Gamalero E, Trotta A, Massa N, Copetta A, Martinotti MG, Berta G. Impact of two fluorescent pseudomonads and an arbuscular mycorrhizal fungus on tomato plant growth, root architecture

and P acquisition. Mycorrhiza. 2004;**14**:185-192

[66] Frey-Klett P, Garbaye J, Tarkka M. The Mycorrhiza helper bacteria revisited. New Phytologist. 2007;**176**:22-36

[67] Poole EJ, Bending GD, Whipps JM, Read DJ. Bacteria associated with *Pinus sylvestris-Lactarius rufus* ectomycorrhizas and their effects on mycorrhiza formation in vitro. New Phytologist. 2001;**151**:743-751

[68] Duponnois R, Garbaye J. Mycorrhizal helper bacteria associated with the Douglas fir *Laccaria laccata* symbiosis: Effects in aseptic and in glasshouse conditions. Annales des Sciences Forestières. 1991;**48**:239-251

[69] Maier A, Riedlinger J, Fiedler HP, Hampp R. Actinomycetales bacteria from a spruce stand: Characterization and effects on growth of root symbiotic and plant parasitic soil fungi in dual culture. Mycological Progress. 2004;**3**(2):129-136

[70] Schrey SD, Schellhammer M, Ecke M, Hampp R, Tarkka MT. Mycorrhiza helper bacterium *Streptomyces* AcH 505 induces differential gene expression in the ectomycorrhizal fungus *Amanita muscaria*. New Phytologist. 2005;**168**:205-216

[71] Riedlinger J, Schrey SD, Tarkka MT, Hampp R, Kapur M, Fiedler HP. Auxofuran, a novel metabolite that stimulates the growth of fly agaric, is produced by the mycorrhiza helper bacterium *Streptomyces* strain AcH 505. Applied and Environmental Microbiology. 2006;**72**:3550-3557

[72] Deveau A, Palin B, Delaruelle C, Peter M, Kohler A, Pierrat JC, et al. The mycorrhiza helper *Pseudomonas fluorescens* BBc6R8 has a specific priming effect on the growth, morphology and gene expression of the ectomycorrhizal fungus *Laccaria bicolor* S238N. New Phytologist. 2007;**175**:743-755

[73] Zhou AD, Wu XQ, Shen L, Xu XL, Huang L, Ye JR. Profiling of differentially expressed genes in ectomycorrhizal fungus *Pisolithus tinctorius* responding to mycorrhiza helper *Brevibacillus reuszeri* MPt17. Biologia. 2014;**69**(4):435-442

[74] Artursson V, Finlay RD, Jansson JK. Interactions between arbuscular mycorrhizal fungi and bacteria and their potential for stimulating plant growth. Environmental Microbiology. 2006;**8**:1-10

[75] Avis TJ, Gravel V, Autoun H, Tweddel RJ. Multifaceted beneficial effects of rhizosphere microorganisms on plant health and productivity. Soil Biology and Biochemistry. 2008;**40**:1733-1740

[76] Gravel V, Antoun H, Tweddell RJ. Growth stimulation and growth yield improvement of greenhouse tomato plants by inoculation with *Pseudomonas putida* and *Trichoderma atroviride*: Possible role of indole acetic acid (IAA). Soil Biology and Biochemistry. 2007;**39**:1968-1977

[77] Hayat R, Ali S, Amara U, Khalid R, Ahmed I. Soil beneficial bacteria and their role in plant growth promotion: A review. Annales de Microbiologie. 2010;**60**:579-598

[78] Malusa E, Sas-Paszt L, Ciesielska J. Technologies for beneficial microorganisms inocula used as biofertilizers. The Scientific World Journal. 2012;**2012**:12. DOI: 10.1100/2012/491206. Article ID: 491206

[79] Vestberg M, Kukkonen S, Saari K, Parikka P, Huttunen J, Tainio L, et al.

Microbial inoculation for improving the growth and health of micropropagated strawberry. Applied Soil Ecology. 2004;**27**:243-258

[80] Azcón-Aguilar C, Barea JM, Gianinazzi S, Gianinazzi-Pearson V, editors. Mycorrhizas Functional Processes and Ecological Impact. Berlin: Springer; 2009

[81] Barea JM, Azcón R, Azcón-Aguilar C. Interactions between mycorrhizal fungi and bacteria to improve plant nutrient cycling and soil structure. In: Varma A, Buscot F, editors. Microorganisms in Soils: Roles in Genesis and Functions, Vol. 3. Heidelberg: Springer Berlin; 2005. pp. 195-212

[82] Hata S, Kobae Y, Banba M. Interactions between plants and arbuscular mycorrhizal fungi. In: Kwang WJ, editor. International Review of Cell and Molecular Biology, Vol. 281. San Diego, USA: Academic Press; 2010. pp. 1-48

[83] Reddy CA, Saravanan RS. Polymicrobial multi-functional approach for enhancement of crop productivity. In: Sariaslani S, Gadd GM, editors. Advances in Applied Microbiology. Vol. 82. Michigan, USA: Elsevier; 2013. pp. 53-113

[84] Andre S, Galiana A, Le Roux C, Prin Y, Neyra M, Duponnois R. Ectomycorrhizal symbiosis enhanced the efficiency of inoculation with two *Bradyrhizobium* strains and *Acacia holosericea* growth. Mycorrhiza. 2005;**15**:357-364

[85] Saravanan RS, Natarajan K. Effect of *Pisolithus tinctorius* on the nodulation and nitrogen fixing potential of *Acacia nilotica* seedlings. Kavaka. 1996;**24**:41-49

[86] Saravanan RS, Natarajan K. Effect of ecto- and endomycorrhizal fungi along with *Bradyrhizobium sp*. on the growth and nitrogen fixation in *Acacia nilotica* seedlings in the nursery. Journal of Tropical Forest Science. 2000;**12**:348-356

[87] Seneviratne G, Zavahir J, Bandara W, Weerasekara M. Fungal-bacterial biofilms: Their development for novel biotechnological applications. World Journal of Microbiology and Biotechnology. 2008;**24**:739-743

[88] Ude S, Arnold DL, Moon CD, Timms-Wilson T, Spiers AJ. Biofilm formation and cellulose expression among diverse environmental *Pseudomonas* isolates. Environmental Microbiology. 2006;**8**:1997-2011

[89] Garbaye J. Use of mycorrhizas in forestry. In: Strullu DG, editor. Les mycorhizes des Arbres et Plantes Cultivées. Paris, France: Lavoisier; 1990. pp. 197-248

[90] DuñabeitiaM, RodríguezN, SalcedoI, Sarrionandia E. Field mycorrhization and its influence on the establishment and development of the seedlings in a broadleaf plantation in the Basque country. Forest Ecology and Management. 2004;**195**:129-139

[91] Dahlberg A, Genney DR, Heilmann-Clausen J. Developing a comprehensive strategy for fungal conservation in Europe: Current status and future needs. Fungal Ecology. 2010;**3**(2):50-64

[92] McAfee BJ, Fortin JA. Competitive interactions of ectomycorrhizal mycobionts under field conditions. Canadian Journal of Botany. 1986;**64**:848-852

[93] Domínguez JA, Selva J, Rodríguez Barreal JA, de Omeñaca S. The influence of mycorrhization with *Tuber melanosporum* in the afforestation of a Mediterranean site with *Quercus ilex* and *Quercus faginea*. Forest Ecology and Management. 2006;**231**:226-233

[94] Olivier JM, Savignac JC, Sourzat P. Truffe et Trufficulture. Perigueux, France: Ed Fanlac; 1996

[95] Tibbett M, Cairney JWG. The cooler side of mycorrhizas: Their occurrence and functioning at low temperatures. Canadian Journal of Botany. 2007;**85**:51-62

[96] Geml J, Timling I, Robinson CH, Lennon N, Nusbaum HC, Brochmann C, et al. Antarctic community of symbiotic fungi assembled by long-distance dispersers: Phylogenetic diversity of ectomycorrhizal basidiomycetes in Svalbard based on soil and sporocarp DNA. Journal of Biogeography. 2011;**39**:74-88

[97] Guerrero-Galán C, Calvo-Polanco M, Zimmermann SD. Ectomycorrhizal symbiosis helps plants to challenge salt stress conditions. Mycorrhiza. 2019;**29**:291-301. DOI: 10.1007/s00572-019-00894-2

[98] Colpaert JV, Wevers JHL, Krznaric E, Adriaensen K. How metal-tolerant ecotypes of ectomycorrhizal fungi protect plants from heavy metal pollution. Annals of Forest Science. 2011;**68**:17-24

[99] Azul AM, Sousa JP, Agerer R, Martín MP, Freitas H. Land use practices and ectomycorrhizal fungal communities from oak woodlands dominated by *Quercus suber* L. considering drought scenarios. Mycorrhiza. 2010;**20**:73-88

[100] Blom JM, Vannini A, Vettraino AM, Hale MD, Godbold DL. Ectomycorrhizal community structure in a healthy and a *Phytophthora*-infected chestnut (*Castanea sativa* Mill.) stand in central Italy. Mycorrhiza. 2009;**20**:25-38. DOI: 10.1007/s00572-009-0256-z

[101] Branzanti MB, Rocca E, Pisi A. Effect of ectomycorrhizal fungi on chestnut ink disease. Mycorrhiza. 1999;**9**:103-109. DOI: 10.1007/s005720050007

[102] Corcobado T, Moreno G, Azul AM, Solla A. Seasonal variations of ectomycorrhizal communities in declining *Quercus ilex* forests: Interactions with topography, tree health status and *Phytophthora cinnamomi* infections. Forestry. 2015;**88**:257-266. DOI: 10.1093/forestry/cpu056

[103] Lehto T, Zwiazek JJ. Ectomycorrhizas and water relations of trees: A review. Mycorrhiza. 2011;**21**(2):71-90. DOI: 10.1007/s00572-010-0348-9

[104] Maurel C, Plassard C. Aquaporins: For more than water at the plant-fungus interface? New Phytologist. 2011;**190**(4):815-817. DOI: 10.1111/j.1469-8137.2011.03731.x

[105] Nehls U, Dietz S. Fungal aquaporins: Cellular functions and ecophysiological perspectives. Fungal Aquaporins: Cellular functions and Ecophysiological perspectives. Applied Microbiology and Biotechnology. 2014;**98**(21):8835-8851. DOI: 10.1007/s00253-014-6049-0

[106] Xu H, Kemppainen M, El Kayal W, Lee SH, Pardo AG, Cooke JEK, et al. Overexpression of *Laccaria bicolor* aquaporin JQ585595 alters root water transport properties in ectomycorrhizal white spruce (*Picea glauca*) seedlings. New Phytologist. 2015;**205**:757-770. DOI: 10.1111/nph.13098

[107] Ishida TA, Nara K, Ma S, Takano T, Liu S. Ectomycorrhizal fungal community in alkaline-saline soil in northeastern China. Mycorrhiza. 2009;**19**(5):329-335. DOI: 10.1007/s00572-008-0219-9

[108] Luo ZB, Li K, Gai Y, Gobel C, Wildhagen H, Jiang XN, et al. The ectomycorrhizal fungus (*Paxillus*

involutus) modulates leaf physiology of poplar towards improved salt tolerance. Environmental and Experimental Botany. 2011;**72**:304-311. DOI: 10.1016/j.envexpbot.2011.04.008

[109] Richard F, Roy M, Shahin O, Sthultz C, Duchemin M, Joffre R, et al. Ectomycorrhizal communities in a Mediterranean forest ecosystem dominated by *Quercus ilex*: Seasonal dynamics and response to drought in the surface organic horizon. Annals of Forest Science. 2011;**68**:57-68. DOI: 10.1007/s13595-010-0007-5

[110] Li J, Bao S, Zhang Y, Ma X, Mishra-Knyrim M, Sun J, et al. *Paxillus involutus* strains MAJ and NAU mediate K(+)/Na(+) homeostasis in ectomycorrhizal *Populus x canescens* under sodium chloride stress. Plant Physiology. 2012;**159**:1771-1786. DOI: 10.1104 /pp.112.195370

[111] Luo ZB, Janz D, Jiang X, Göbel C, Wildhagen H, Tan Y, et al. Upgrading root physiology for stress tolerance by ectomycorrhizas: Insights from metabolite and transcriptional profiling into reprogramming for stress anticipation. Plant Physiology. 2009;**151**:1902-1917. DOI: 10.1104/pp.109.143735

[112] Szuba A. Ectomycorrhiza of *Populus*. Forest Ecology and Management. 2015;**347**:156-169. DOI: 10.1016/ j.foreco.2015.03.012

[113] Beaudoin-Nadeau M, Gagné A, Bissonnette C, Bélanger P, Fortin J, Roy S, et al. Performance of ectomycorrhizal alders exposed to specific Canadian oil sands tailing stressors under in vivo bipartite symbiotic conditions. Canadian Journal of Microbiology. 2016;**62**(7):543-549. DOI: 10.1139/cjm-2015-0703

[114] Adriaensen K, van der Lelie D, Van Laere A, Vangronsveld J, Colpaert JV. A zinc-adapted fungus protects pines from zinc stress. New Phytologist. 2004;**161**:549-555. DOI: 10.1046/j.1469-8137.2003.00941.x

[115] Adriaensen K, Vrålstad T, Noben JP, Vangronsveld J, Colpaert JV. Copper-adapted *Suillus luteus*, a symbiotic solution for pines colonizing Cu mine spoils. Applied and Environmental Microbiology. 2005;**71**:7279-7284. DOI: 10.1128/AEM.71.11.7279-7284.2005

[116] Jourand P, Ducousso M, Reid R, Majorel C, Richert C, Riss J, et al. Nickel-tolerant ectomycorrhizal *Pisolithus albus* ultramafic ecotype isolated from nickel mines in New Caledonia strongly enhance growth of the host plant *Eucalyptus globulus* at toxic nickel concentrations. Tree Physiology. 2010;**30**:1311-1319. DOI: 10.1093/treephys/tpq070

[117] Kayama M, Yamanaka T. Growth characteristics of ectomycorrhizal seedlings of *Quercus glauca*, *Quercus salicina*, and *Castanopsis cuspidata* planted on acidic soil. Trees. 2014;**28**:569-583. DOI: 10.1007/s00468-013-0973-y

[118] Luo ZB, Wua C, Zhang C, Lic H, Lipkad U, Polle A. The role of ectomycorrhizas in heavy metal stress tolerance of host plants. Environmental and Experimental Botany. 2014;**108**:47-62. DOI: 10.1016/j.envexpbot.2013.10.018

[119] Meharg AA, Cairney JWG. Ectomycorrhizas: Extending the capacities of rhizosphere remediation? Soil Biology and Biochemistry. 2000;**32**:1475-1484

[120] Krupa P, Kozdrój J. Accumulation of heavy metals by ectomycorrhizal fungi colonizing birch trees growing in an industrial desert soil. World Journal of Microbiology and Biotechnology. 2004;**20**(4):427-430. DOI: 10.1023/B:WIBI.0000033067.64061.f3

[121] Machuca A. Metal-chelating agents from ectomycorrhizal fungi and their biotechnological potential. In: Rai M, Varma A, editors. Diversity and Biotechnology of Ectomycorrhizas. Soil Biology. Berlin, Heidelberg: Springer-Verlag; 2011:**25**

[122] Daghino S, Martino E, Perotto S. Model systems to unravel the molecular mechanisms of heavy metal tolerance in the ericoid mycorrhizal symbiosis. Mycorrhiza. 2016;**26**:263-274. DOI: 10.1007/s00572-015-0675-y

[123] Krpata D, Fitz W, Peintner U, Langer I, Schweiger P. Bioconcentration of zinc and cadmium in ectomycorrhizal fungi and associated aspen trees as affected by level of pollution. Environmental Pollution. 2009;**157**:280-286. DOI: 10.1016/j.envpol.2008.06.038

Chapter 9

Microbes for Iron Chlorosis Remediation in Peach

Saurabh Kumar Singh

Abstract

Peach [*Prunus persica* (L.) Batsch] suffers from iron chlorosis when grown in calcareous soils due to low iron availability. Traditionally, soil and foliar application of ferrous sulphate, Fe-EDTA, Fe-EDDHA chelates, etc. is used as a corrective measure of chlorosis. The latter practice is quite effective. However, variable responses have been reported. Therefore, foliar spray cannot yet be considerd as a reliable method to control lime-induced chlorosis. Bioremediation constitutes innovative approaches for chlorosis correction. Iron fixations in calcareous soil, iron uptake by plants, and advance detection techniques and correction strategies in plants for iron chlorosis have been discussed in this chapter. The microbe-mediated correction strategies are identified as eco-friendly.

Keywords: peach, *Prunus persica*, calcareous soil, iron chlorosis, bioremediation

1. Introduction

Peach [*Prunus persica* (L.) Batsch] is one of the most common temperate region fruit crops of the world. China, Italy, the USA, Greece, Spain, Turkey, Iran, Chile, etc. are the major producing countries [1]. This stone fruit crop belongs to the family Rosaceae. Peach [*Prunus persica* var. *vulgaris* Maxim.] with round and fuzzy fruit, the nectarine [*Prunus persica var. nectarina* (Aiton) Maxim.] with round fruit but without pubescence (fuzz), and the flat peach [*Prunus persica* var. *platicarpa* Bailey] with flat-shaped fruit are the three categories [2]. Iron, the fourth most prevalent element preceded by O, Si, and Al in the earth's crust and soils, is classified as an essential micronutrient for plant growth. It is a multifunctional element [3], required for the different physicochemical processes of plants, and plays an important role in chlorophyll activation, chloroplast membrane structure, photosynthesis, respiration, and synthesis of many heme proteins and iron–sulphur (Fe-S) clusters as cofactors of proteins that function in the fundamental life of plants [4–6]. Higher plants use two general mechanisms (strategies I and II) for iron acquisition with low iron availability in soil [7]. Calcareous soil gives lower iron availability abreast with a diminishing uptake efficiency by plant roots specially of a plant that depends on ferric reductase activity, because of higher soil pH and bicarbonate concentration [8, 9]. Out of a total of 13.4 billion ha global land surface, 1.5 billion ha is used in crop production, including arable lands plus lands under permanent crops [10, 11]. 30% of the soils in the world are calcareous in nature. They limit the iron availability for plant growth and development, not due to the iron status of the soil but due to their solubility [12].

Causes	Detection methods	Control measures	
		Traditional	Advanced
1. Calcareous soil 2. High soil pH 3. Low chelating ability of roots 4. Lower translocation of iron in plant system 5. Fast vegetative growth 6. High bicarbonate in soil	1. Visual symptoms as intervenous chlorosis 2. Advance detection via physiological and molecular markers	1. Application of iron sources 2. Soil application 3. Foliar application 4. Trunk injection	1. Bioremediation 2. Application of nanofertilizers 3. Transgenic breeding approach

Table 1.
An overview of the causes, detection methods, and control measures of iron chlorosis in peach.

Iron chlorosis in calcareous soils is often termed as lime-induced iron chlorosis [13]. Applications of iron sources either in soil or as foliar spray are generally practised to correct the iron chlorosis in peach. An overview of the causes, detection methods, and control measures is given in **Table 1**. Kloepper et al. [14] gave a pioneer verification of iron-depriving microflora in soil and reported the plant growth-promoting activity of rhizobacteria pertaining to the iron-chelating siderophores. These are low-molecular-weight metabolites having a high affinity for Fe(III). Involvement of siderophore and proton production resulted in improved iron bioavailability in the root zone of plants [15, 16]. This chapter addresses the current trend of detection methods and control measures of iron cholorosis in peach and gives attention to bioremediation techniques for the correction of lime-induced iron cholorosis.

2. Iron fixation in calcareous soil

Calcareous soils have often more than 15% $CaCO_3$. Soil with high $CaCO_3$ belongs to calcisols and related calcic subgroup of other soils, dominantly found in dried areas of the earth [17]. Plants show iron stress when grown in calcareous soil due to lower concentration of available iron [18]. The two oxidation states of iron are the reduced form, i.e. ferrous iron (Fe^{2+}), and the oxidized form, i.e. ferric iron (Fe^{3+}), in all living forms. $CaCO_3$ directly participates in the reactions that decrease the iron availability to the plants. The reactions of iron fixation are as follows:

$$Fe^{2+} + CaCO_3 \rightleftharpoons FeCO_3 + Ca^{2+} \tag{1}$$

$$4FeCO_3 + O_2 + Ca(HCO_3)_2 \rightleftharpoons 2Fe_2(CO_3)_2 + Ca(OH)_2 \tag{2}$$

$$Fe_2(CO_3)_3 + 3H_2O \rightleftharpoons Fe_2O_3 + 3H_2CO_3 \tag{3}$$

Ferrous iron (Fe^{2+}) is fixed as ferric oxide (Fe_2O_3) and becomes unavailable to plant roots.

3. Mechanism for iron uptake in higher plants

The iron uptake mechanisms of higher plants can be categorized into two groups as plant strategy and microbe mediated. In plant strategy mechanism, plants use two strategies, viz. strategy I and strategy II of iron uptake [19], whereas in microbe mediated through Fe siderophore complexes [14]. A brief description of uptake mechanisms is given in the following subheadings.

3.1 Plant strategies for iron uptake

From small seasonal cereal crops like rice, wheat, etc. to the perennial tall fruit crops, two strategies are recognized for iron uptake: strategy I (for dicots and nongraminaceous monocots) and strategy II (for graminaceous species) [7, 8, 20, 21].

Dicots and nongraminaceous monocots use strategy I for iron destabilization in the root zone of the plants. The reduction of Fe^{3+} to Fe^{2+} at the root surface, increased proton (H^+) extrusion, and release of reducing and/or chelating substances are the three main mechanisms in the plants that use strategy I [20], whereas strategy II is expressed only in the grass family. Exudation of iron-chelating compounds, i.e. phytosiderophores (non-proteinogenic amino acids), from the roots helps in mobilizing Fe(III) as Fe phytosiderophore complexes. Finally, the Fe phytosiderophore complexes are absorbed by plant roots [7]. Peach suffers from iron chlorosis due to lower efficiency of iron chelation at the root zone in calcareous soils. The ferric-chelate reductase (FC-R) ability of the roots can be used for Fe^{3+} tolerance screening tool [22].

3.2 Microbe-mediated iron uptake

Besides strategies I and II of the plants to absorb iron under limiting conditions, there is also microbial solubilization of iron in the soil. Evidence of plant growth-promoting rhizobacteria (PGPR)-mediated iron bio-solubilization was reported by Kloepper et al. [14]. A number of microbes that predominantly belong to *Pseudomonas* and *Trichoderma* genera of bacterial and fungal groups, respectively, have been reported for bio-solubilization of iron. They release siderophore, like the phytosiderophores of the plants of strategy II group. Siderophores are low-molecular-weight (500–1500 daltons) iron-chelating compounds [15], synthesized by micro-organisms, i.e. *Pseudomonas*, *Azotobacter*, *Bacillus*, *Enterobacter*, *Serratia*, *Azospirillum*, *Rhizobium*, *Trichoderma*, *Cenococcum geophilum*, and *Suillus granulatus* [23–29]. Microbial siderophores are structurally diverse low-molecular-mass (200–2000 Da) [30, 31] compounds, with distinctive characteristics of Fe siderophore complex formation. Siderophores are usually classified by the ligands used to chelate the ferric iron by moieties donating the oxygen ligands for Fe(III) coordination and its specific chemical property. The major groups of siderophores include the catecholates, hydroxamates, and carboxylates. The catecholate is a dominating siderophore produced by bacteria, whereas the hydroxamate is produced by fungi [15, 32, 33]. They make stable complex with iron as Fe siderophore soluble complex, in soil solution and at the mineral surface, and then become available for uptake by the cell membrane of plant roots. Further, upon absorption, siderophores of Fe siderophore complexes are either recycled or destroyed [34–36]. Due to complex formation property of siderophores with iron, the Fe siderophore form of soil iron which can be utilized in controlling chlorosis of peach grown in calcareous soil has been little explored hitherto.

4. Markers for advance detection of Fe chlorosis

Chlorophyll content [37, 38], SPAD index [38–42], chlorophyll fluorescence [43, 44], thylakoid membrane lipids [45], photosynthetic rate [46], physiologically active iron [47–50], Fe/Mn ratio [51, 52], and transformed reflectance spectra [53] are important physiological parameters used for the detection of iron chlorosis in different crops. Literature supports the possibility of using physiological and molecular markers as advance detection technique of iron chlorosis.

4.1 Physiological markers

Brown [54] emphasized to study the biochemical basis of iron chlorosis and its contributing factors. Efficiency of iron uptake depends on plant species [55]. Iron status of different plant parts like leaves, bark, flowers, vegetative buds, and floral buds has been reported by using tissue index in different crops for predicting the iron chlorosis. Floral analysis is reported as a tool for prediction of iron deficiency in peach [56, 57].

Iron plays an important role in chlorophyll formation [58, 59]. The reduction in the number of granal and stromal lamellae per chloroplast and in the number of thylakoids per granum under iron stress condition was reported by Spiller and Terry [60]. In parallel, Terry [61] also reported a decrease in chlorophyll (Chl) a and Chl b contents of sugar beet (*Beta vulgaris* L.) leaves under Fe stress condition and no effect on the number of chloroplast per unit area. The quantitative reduction (75%) in chlorophyll content per unit area and role of iron in chloroplast development were also noted in sugar beet [62, 63]. The findings showed that there is a quantitative decrease in chlorophyll content of leaves under iron stress condition. Chlorophyll fluorescence and iron concentration in the flowers of peach, root apoplastic iron in soya bean, and morphological changes of plant root coupled with alteration in citrate concentration in the phloem of castor bean are found to be directly correlated with chlorosis under iron stress condition [64–67]. So, chlorophyll content of leaves, SPAD index reading, chlorophyll fluorescence, concentration of iron in plant parts, and change in root morphology can be used as markers for advance detection of iron stress. These predictions will be helpful in managing the iron chlorosis in peach. Foliar iron application could be used for remediation of chlorosis problem [68]. Nicotianamine (a non-proteinogenic amino acid), nitric oxide levels, and concentration of nutrients in the reproductive buds need extensive research to be used as markers for the selection of Fe efficient genotypes in *Prunus* sp. [69–73].

4.2 Molecular markers

The need to search for the blueprint of iron transport, molecular mechanism of genes controlling iron uptake, and intracellular storage was emphasized by Briat and Lobreaux [74]. Current researches clarified that in different micro-organisms, a small regulatory RNA, RyhB, plays an essential role in the metabolism of iron. Numerous data on the molecular level of iron transport in plants are published, and there is a need for a comprehensive research on iron homeostasis [19, 75, 76]. *Arabidopsis thaliana* (arabidopsis), *Lycopersicon esculentum* (tomato), and *Pisum sativum* (pea) are used as model plants to study strategy I of iron absorption. Iron is translocated as a ferric citrate complex in the xylem with the help of FRD3 effluxes of citrate, from root to shoot portion of plants [77]. A lot of information for molecular basis of iron transport and compartment have been decoded. There is a need to spell out each Fe translocation step, iron chelator complex, Fe flux, signal, and receptor regulating the Fe nutritional status [78, 79].

Fe is concentrated in the vacuoles of cells. A group of co-expression genes is involved in iron deficiency regulation [80, 81]. In iron translocation, there are functional links between Fe loading in vacuoles (AtVIT1 gene) and remobilization (AtVIT1 and AtNRAMP3 genes) in arabidopsis, and iron accumulation in vacuolar globoids is obstructed with mystifying genes [80]. Gonzalo et al. [82] studied P 2175 (myrobalan plum) and Felinem (peach-almond hybrid) for the differential expression of genes involved in homeostasis. The genes PFRO2 (for reductase activity), PIRT1 (for transport in roots), and PAHA2 (for proton release) were expressed, and can be used as molecular markers in screening and developing cultivars as well as the rootstocks of fruit crops for iron tolerance. Molecular advancement of iron regulation and decoding of iron regulatory gene will be helpful in managing iron chlorosis in peach.

4.3 Index tissue

Based on nutrient status of a specific plant part, the fertilizer rate may be recommended to correct the nutritional deficiency. In sampling, the age of selected plant part and time of sampling should be considered. Concurrently, the sampling of plants damaged due to insect pest infestation, pathogen attacks, and mechanical injuries must be avoided [83]. Details in plant analysis principles, sampling procedure, and laboratory analysis are given by Jones [83]. In peach, the leaves near the current year growth should be sampled during mid-season of growth, with a sample size of 50–100 selected plants. The best sampling time in peach with correlation to yield was found at 60 days after full bloom [84].

5. Microbes for iron chlorosis remediation

Soil and foliar application of synthetic iron sources is used for controlling iron chlorosis in peach. The latter practice is quite effective. Foliar Fe fertilization is a widespread agricultural strategy to control lime-induced iron chlorosis [68, 85]. However, variable responses to Fe sprays have often been described, and foliar Fe fertilization cannot yet be considered a reliable strategy to control plant Fe deficiency [86, 87]. Soil applications of iron sources have their own limitation. Due to its oxidized form as ferric state in soil, it forms very insoluble minerals. In addition to their practical applicability and intricacy, these chelated chemicals are too expensive. Due to the limitation of application of iron source, microorganism-mediated bioavailability of iron can be an effective way to control iron chlorosis.

Crowley et al. [88] confer the existence of a microbial siderophore iron transport system in oat (*Avena sativa* cv. Victory). Application of bacterial siderophore of the two siderophore-producing bacterial strains, namely, *Chryseobacterium* spp. C138 from the rhizosphere of *Oryza sativa* and *Pseudomonas fluorescens* N21.4 from the rhizosphere of *Nicotiana glauca*, in iron-starved tomato plants grown in hydroponic culture resulted significantly in higher plant yield and chlorophyll and iron content [89]. The findings indicated that siderophores are helpful in providing iron to plant. Another experiment on red bean under greenhouse condition showed an increase in bean plant growth factors, significantly inoculated with 7NSK2, UTPF5, and UTPF 76 strains of fluorescent *Pseudomonas* [90]. The beneficial effects of microbial siderophores have potential to correct lime-induced chlorosis in peach.

6. Conclusions

Peach is unexplored in terms of application of bioremediation. It is therefore necessary to evaluate the response of microorganism for controlling iron chlorosis in peach, grown in calcareous soils. Microbial iron mobilization needs vast research for identifying efficient strains regarding iron mobilization and their effect on plant growth, nutritional status, and yield. Bioremediation will help in reducing the dependency on chemical measure of controlling chlorosis in addition to eco-friendly remediation as a long-term solution.

Author details

Saurabh Kumar Singh
Department of Horticulture, School of Agriculture, Lovely Professional University, Phagwara, Punjab, India

*Address all correspondence to: saurabh3596@gmail.com

References

[1] FAOSTAT. 2012. Available from: http://faostat.fao.org/site/339/default.aspx searched [Accessed: 13 February 2016]

[2] LaReu JH. Introduction. In: LaReu HH, Johnson RS, editors. Peaches, Plums and Nectarines. Oakland: University of California; 1989

[3] Briat JF, Curie C, Gaymard F. Iron utilization and metabolism in plants. Current Opinion in Plant Biology. 2007;**10**:276-282

[4] Monge E, Perez C, Pequerul A, Madero P, Val J. Effect of iron chlorosis on mineral nutrition and lipid composition of thylakoid biomembrane in *Prunus persica* (L.) Bastch. Plant and Soil. 1993;**154**:97-102

[5] Connolly EL, Guerinot ML. Iron stress in plants. Genome Biology. 2002;**3**:1024.1-1024.4

[6] Balk J, Pilon M. Ancient and essential: The assembly of iron–sulfur clusters in plants. Trends in Plant Science. 2011;**16**:218-226

[7] Marschner H, Romheld V, Kissel M. Different strategies in higher plants in mobilization and uptake of iron. Journal of Plant Nutrition. 1986;**9**:695-713

[8] Jeong J, Connolly EL. Iron uptake mechanisms in plants: Functions of the FRO family of ferric reductases. Plant Science. 2009;**176**:709-714

[9] Lucena C, Romera FJ, Garcia MJ, Alcantara E, Perez-Vicente R. Ethylene participates in the regulation of Fe deficiency responses in strategy I plants and in rice. Frontiers in Plant Science. 2015;**6**:1056

[10] World Food and Agriculture. FAO Statistical Yearbook-2013. Food and Agriculture Organization of United Nations; 2013

[11] FAO Corporate Document Repository. Crop Production and Natural Resource Use. 2006. Available from: http://www.fao.org/docrep/005/y4252e/y4252e06.htm

[12] Guerinot ML. Improving rice yields – Ironing out the details. Nature Biotechnology. 2001;**19**:417-418

[13] Pestana M, Varennes AD, Faria EA. Diagnosis and correction of iron chlorosis in fruit trees: A review. Journal of Food, Agriculture and Environment. 2003;**1**:46-51

[14] Kloepper JW, Leong J, Teintze M, Shroth MN. Enhanced plant growth by siderophores produced by plant growth promoting rhizobacteria. Nature. 1980;**286**:885-886

[15] Hider RC, Kong X. Chemistry and biology of siderophores. Natural Product Reports. 2010;**27**:637-657

[16] Jin CW, Ye YQ, Zheng SJ. An underground tail: Contribution of microbial activity to plant iron acquisition via ecological processes. Annals of Botany. 2014;**113**:7-18

[17] Management of Calcareous Soils. FAO Soil Portal, Food and Agriculture Organization of United Nations. Available from: http://www.fao.org/soils-portal/soil-management/management-of-some-problem-soils/calcareous-soils/en/ [Accessed: 22 January 2016]

[18] Loeppert RH. Reactions of iron and carbonates in calcareous soils. Journal of Plant Nutrition. 1986;**9**:3-7

[19] Hell R, Stephan UW. Iron uptake, trafficking and homeostasis in plants. Planta. 2003;**216**:541-551

[20] Romheld V. Different strategies for iron acquisition in higher plants. Physiologia Plantarum. 1987;**70**:231-234

[21] Marschner H, Romheld V. Strategies of plants for acquisition of iron. Plant and Soil. 1994;**165**:261-274

[22] Gogorcena Y, Abadia J, Abadia A. A new technique for screening iron-efficient genotypes in peach rootstocks: Elicitation of root ferric chelate reductase by manipulation of external iron concentrations. Journal of Plant Nutrition. 2005;**27**:1701-1715

[23] Meyer JM, Abdallah MA. The fluorescent pigment of pseudomonas fluorescens: Biosynthesis, purification and physiochemical properties. Microbiology. 1978;**107**:319-328

[24] Loper JE, Buyer JS. Siderophores in microbial interactions on plant surfaces. Molecular Plant-Microbe Interactions. 1991;**4**:5-13

[25] Watteau F, Berthelin J. Microbial dissolution of iron and aluminum from soil minerals – Efficiency and specificity of hydroxamate siderophores compared to aliphatic acids. European Journal of Soil Biology. 1994;**30**:1-9

[26] Haselwandter K, Winkelmann G. Ferricrocin—An ectomycorrhizal siderophore of *Cenococcum geophilum*. Biometals. 2002;**15**:73-77

[27] Ali SS, Vidhale NN. Bacterial siderophore and their application: Review. International Journal of Current Microbiology and Applied Sciences. 2013;**2**(12):303-312

[28] Winkelmann G. Ecology of siderophores with special reference to the fungi. Biometals. 2007;**20**:379-392

[29] Vinale F, Nigro M, Sivasithamparam K, Flematti G, Ghisalberti EL, Ruocco M, et al. Harzianic acid: A novel siderophore from *Trichoderma harzianum*. FEMS Microbiology Letters. 2013;**347**:123-129

[30] Schwyn B, Neilands JB. Universal chemical assay for the detection and determination of siderophores. Analytical Biochemistry. 1987;**160**:47-56

[31] Wang W, Qiu Z, Hongming T, Cao L. Siderophore production by actinobacteria. Biometals. 2014;**27**:623-631

[32] Miethke M, Marahiel MA. Siderophore based iron acquisition and pathogen control. Microbiology and Molecular Biology Reviews. 2007;**71**:413-451

[33] Schalk IJ, Hannauer M, Braud A. New roles for bacterial siderophores in metal transport and tolerance. Environmental Microbiology. 2011;**13**:2844-2854

[34] Crowley DE, Wang YC, Reid CPP, Szaniszlo PJ. Mechanism of iron acquisition from siderophore by microorganism and plants. Plant and Soil. 1991;**130**:179-198

[35] Kraemer SM. Iron oxide dissolution and solubility in the presence of siderophores. Aquatic Sciences. 2004;**66**:3-18

[36] Kraemer SM, Crowley DE, Kretzschmar R. Geochemical aspects of phytosiderophore-promoted iron acquisition by plants. Advances in Agronomy. 2006;**91**:1-46

[37] Marsh JHV, Evans HJ, Matrone G. Investigations of the role of iron in chlorophyll metabolism I. Effect of iron deficiency on chlorophyll and heme content and on the activities of certain enzymes in leaves. Plant Physiology. 1963;**38**:632

[38] Smith BR, Fisher PR, Argo WR. Growth and pigment content of container-grown impatiens and petunia in relation to root substrate pH and applied micronutrient concentration. HortScience. 2004;**39**:1421-1425

[39] Peryea FJ, Kammereck R. Use of Minolta SPAD-502 chlorophyll meter to quantify the effectiveness of mid-summer trunk injection of iron on chlorotic pear trees. Journal of Plant Nutrition. 1997;**20**:1457-1463

[40] Fisher PR, Wik RM, Smith BR, Pasian CC, Kmetz-González M, Argo WR. Correcting iron deficiency in calibrachoa grown in a container medium at high pH. HortTechnology. 2003;**13**:308-313

[41] Cinelli F, Loreti F. Evaluation of some plum rootstocks in relation to lime-induced chlorosis by hydroponic culture. Acta Horticulturae. 2004;**658**:421-427

[42] Jimenez S, Pinochet J, Abadia A, Moreno MA, Gogorcena Y. Tolerance response to iron chlorosis of Prunus selections as rootstocks. HortScience. 2008;**43**:304-309

[43] Morales F, Cerovic ZG, Moya I. Characterization of blue-green fluorescence in the mesophyll of sugar beet (*Beta vulgaris* L.) leaves affected by iron deficiency. Plant Physiology. 1994;**106**:127-133

[44] Lichtenthaler HK, Miehe JA. Fluorescence imaging as a diagnostic tool for plant stress. Trends in Plant Science. 1997;**2**:316-320

[45] Nishio JN, Taylor SE, Terry N. Changes in thylakoid galactolipids and proteins during iron nutrition-mediated chloroplast development. Plant Physiology. 1985;**77**:705-711

[46] Terry N. Limiting factors in photosynthesis IV. Iron stress-mediated changes in light-harvesting and electron transport capacity and its effects on photosynthesis in vivo. Plant Physiology. 1983;**71**:855-860

[47] Oserkowsky J. Quantitative relation between chlorophyll and iron in green and chlorotic pear leaves. Plant Physiology. 1933;**8**:449

[48] Jacobson L. Iron in the leaves and chloroplasts of some plants in relation to their chlorophyll content. Plant Physiology. 1945;**20**:233

[49] Katyal JC, Sharma BD. A new technique of plant analysis to resolve iron chlorosis. Plant and Soil. 1980;**55**:105-119

[50] Koseoglu AT, Acikgoz V. Determination of iron chlorosis with extractable iron analysis in peach leaves. Journal of Plant Nutrition. 1995;**18**:153-161

[51] Somers II, Gilbert SG, Shive JW. The iron-manganese ratio in relation to the respiratory CO_2 and deficiency-toxicity symptoms in soybeans. Plant Physiology. 1942;**17**:317

[52] Alvarez-Fernandez A, Garcia-Marco S, Lucena JJ. Evaluation of synthetic iron(III)-chelates (EDDHA/Fe^{3+}, EDDHMA/Fe^{3+} and the novel EDDHSA/Fe^{3+}) to correct iron chlorosis. European Journal of Agronomy. 2005;**22**:119-130

[53] Li B, Liew OW, Asundi AK. Pre-visual detection of iron and phosphorus deficiency by transformed reflectance spectra. Journal of Photochemistry and Photobiology B: Biology. 2006;**85**:131-139

[54] Brown JC. Iron chlorosis. Annual Review of Plant Physiology. 1956;**7**:171-190

[55] Christ RA. Iron requirement and iron uptake from various iron compounds by different plant species. Plant Physiology. 1974;**54**:582-585

[56] Sanz M, Belkhoja R, Toselli M, Montanes L, Abadia A, Tagliavini M, et al. Floral analysis as a possible tool for the prognosis of iron deficiency in peach. Acta Horticulturae. 1997;**448**:241-246

[57] Sanz M, Pascual J, Machín J. Prognosis and correction of iron chlorosis in peach trees: Influence on fruit quality. Journal of Plant Nutrition. 1997;**20**:1567-1572

[58] Marschner H. Marschner's Mineral Nutrition of Higher Plants. Cambridge, MA: Academic press; 2011

[59] Briat JF, Dubos C, Gaymard F. Iron nutrition, biomass production, and plant product quality. Trends in Plant Science. 2015;**20**:33-40

[60] Spiller S, Norman T. Limiting factors in photosynthesis. Plant Physiology. 1980;**65**:121-125

[61] Terry N. Limiting factors in photosynthesis. Plant Physiology. 1980;**65**:114-120

[62] Taylor SE, Terry N, Huston RP. Limiting factors in photosynthesis. Plant Physiology. 1982;**70**:1541-1543

[63] Nishio JN, Terry N. Iron nutrition-mediated chloroplast development. Plant Physiology. 1983;**71**:688-691

[64] Longnecker N, Welch RM. Accumulation of apoplastic iron in plant roots. Plant Physiology. 1990;**92**:17-22

[65] Val J, Sanz M, Montanes L, Monge E. Application of chlorophyll fluorescence to study iron and manganese defiviencies in peach tree. Acta Horticulturae. 1995;**283**:201-209

[66] Belkhodja R, Morales F, Sanz M, Abadia A, Abadia J. Iron deficiency in peach trees: Effects on leaf chlorophyll and nutrient concentrations in flowers and leaves. Plant and Soil. 1998;**203**:257-268

[67] Schmidke I, Kruger C, Frommichen R, Scholz G, Stephan UW. Phloem loading and transport characteristics of iron in interaction with plant-endogenous ligands in castor bean seedlings. Physiologia Plantarum. 1999;**106**:82-89

[68] Abadia J, Alvarez-Fernandez A, Morales F, Sanz M, Abadia A. Correction of iron chlorosis by foliar sprays. Acta horticulturae. 2002;**594**:115-121

[69] Pich A, Scholz G. Translocation of copper and other micronutrients in tomato plants (*Lycopersicon esculentum* Mill.): Nicotianamine-stimulated copper transport in the xylem. Journal of Experimental Botany. 1996;**47**:41-47

[70] Schmidt W. Mechanisms and regulation of reduction-based iron uptake in plants. The New Phytologist. 1999;**141**:1-26

[71] Douchkov D, Herbik A, Koch G, Mock HP, Melzer M, Stephan UW, et al. Nicotianamine synthase: Gene isolation, gene transfer and application for the manipulation of plant iron assimilation. Plant and Soil. 2002;**241**:115-119

[72] Cellini A, Corpasb FJ, Barrosoc JB, Masiaa A. Nitric oxide content is associated with tolerance to bicarbonate-induced chlorosis in micropropagated Prunus explants. Journal of Plant Physiology. 2011;**168**:1543-1549

[73] El-Jendoubi H, Igartua E, Abadia J, Abadia A. Prognosis of iron chlorosis in pear (*Pyrus communis* L.) and peach (*Prunus persica* L. Batsch) trees using bud, flower and leaf mineral concentrations. Plant and Soil. 2012;**354**:121-139

[74] Briat JF, Lobreaux S. Iron transport and storage in plants. Trends in Plant Science. 1997;**2**:187-193

[75] Masse E, Salvail H, Desnoyers G, Arguin M. Small RNAs controlling iron metabolism. Current Opinion in Microbiology. 2007;**10**:140-145

[76] Curie C, Briat J. Iron transport and signaling in plants. Annual Review of Plant Biology. 2003;**54**:183-206

[77] Durrett TP, Gassmann W, Rogers EE. The FRD3-mediated efflux of citrate into the root vasculature is necessary for efficient iron translocation. Plant Physiology. 2007;**144**:197-205

[78] Kobayashi T, Nishizawa NK. Iron uptake, translocation, and regulation in higher plants. Annual Review of Plant Biology. 2012;**63**:131-152

[79] Balk J, Schaedler TA. Iron cofactor assembly in plants. Annual Review of Plant Biology. 2014;**65**:125-153

[80] Mary V, Ramos MS, Gillet C, Socha A, Giraudat J, Agorio A, et al. Bypassing iron storage in endodermal vacuoles rescues the iron mobilization defect in the *nramp3nramp4* double mutant. Plant Physiology. 2015;**169**:748-759

[81] Li H, Wang L, Yang ZM. Co-expression analysis reveals a group of genes potentially involved in regulation of plant response to iron-deficiency. Gene. 2015;**554**:16-24

[82] Gonzalo MJ, Moreno MA, Gogorcena Y. Physiological responses and differential gene expression in Prunus rootstocks under iron deficiency conditions. Journal of Plant Physiology. 2011;**168**:887-893

[83] Jones JB, Jr. Soil testing and plant analysis: Guides to the fertilization of horticultural crops. In: Janick J, editor. Horticultural Reviews. 2011. DOI: 10.1002/9781118060735.ch1

[84] Sauz M, Herasa L, Montanesa L. Relationships between yield and leaf nutrient contents in peach trees: Early nutritional status diagnosis. Journal of Plant Nutrition. 1992;**15**:1457-1466

[85] Mortvedt JJ. Correcting iron deficiencies in annual and perennial plants: Present technologies and future prospects. Plant and Soil. 1991;**130**:273-279

[86] Tagliavini M, Abadia J, Rombola AD, Abadia A, Tsipouridis C, Marangony B. Agronomic means for the control of iron deficiency chlorosis in deciduous fruit trees. Journal of Plant Nutrition. 2000;**23**:2007-2022

[87] Garcia-Lavina P, Alvarez-Fernandez A, Abadia J, Abadia A. Foliar applications of acids with and without $FeSO_4$ to control iron chlorosis in pear. Acta Horticulturae. 2002;**594**:217-222

[88] Crowley DE, Reid CPP, Szaniszlo PJ. Utilization of microbial siderophores in iron acquisition by oat. Plant Physiology. 1988;**87**:680-685

[89] Radzki W, Manero FJG, Algar E, Garcia JAL, Garcia-Villaraco A, Solano BR. Bacterial siderophores efficiently provide iron to iron starved tomato plants in hydroponics culture. Antonie Van Leeuwenhoek. 2013;**104**:321-330

[90] Omidvari M, Sharifi RA, Ahmadzadeh M, Dahaji PA. Role of fluorescent pseudomonads to increase bean growth factors. The Journal of Agricultural Science. 2010;**2**:242-247